BIOMEMBRANES AND CELL FUNCTION

ANNALS OF THE NEW YORK ACADEMY OF SCIENCES
Volume 414

BIOMEMBRANES AND CELL FUNCTION

Edited by Fred A. Kummerow, Gheorghe Benga, and Ross P. Holmes

The New York Academy of Sciences
New York, New York
1983

Library of Congress Cataloging in Publication Data

Main entry under title:

Biomembranes and cell function.

(Annals of the New York Academy of Sciences; v. 414)
Articles from a workshop entitled, The Role of biomembranes in the integrity and function of cells, held Aug. 3–6, 1982, New York City. Sponsored by the National Science Foundation and the Romanian Council for Science and Technology.
Includes bibliographical references and index.
1. Cell membranes—Congresses. 2. Cell physiology—Congresses. I. Kummerow, Fred A. II. Benga, Gheorghe. III. Holmes, Ross P. IV. National Science Foundation (U.S.) V. Consiliul Național pentru Știință și Tehnologie. VI. Series.
Q11.N5 vol. 414 [QH601] 500s [574.87'5] 83-26879
ISBN 0-89766-222-9
ISBN 0-89766-223-7 (pbk.)

Subject index prepared by Justine Cullinan

Cover: Negative stains of membranes formed by acyltransferase activity. Courtesy of David W. Deamer.

CCP
Printed in the United States of America
ISBN 0-89766-222-9 (cloth)
ISBN 0-89766-223-7 (paper)

ANNALS OF THE NEW YORK ACADEMY OF SCIENCES

VOLUME 414

December 30, 1983

BIOMEMBRANES AND CELL FUNCTION*

Editors and Conference Chairmen
F. A. KUMMEROW, GHEORGHE BENGA, AND ROSS P. HOLMES

CONTENTS

* This volume is the result of a workshop entitled The Role of Biomembranes in the Integrity and Function of Cells, held on August 3–6, 1982 in New York City.

INTRODUCTION

F. A. Kummerow, Gheorghe Benga, and Ross P. Holmes

Department of Food Science
Burnsides Research Laboratory
University of Illinois at Urbana-Champaign
Urbana, Illinois 61801

The articles in this volume stem from a workshop held in New York City in August, 1982 to facilitate scientific exchange between the United States and Romania and between individual scientists. It was sponsored by the National Science Foundation and the Romanian Council for Science and Technology. The participants all have central research interests in the area of membranes and use a variety of experimental approaches to study their structural and functional properties. The aim of the workshop was to expose participants to these different techniques and approaches, to facilitate an understanding of their application, and ultimately to lead to a cross-fertilization of ideas and the development of new approaches. To this extent, the workshop was highly successful and a rewarding experience for all the participants. It is our belief that readers of the articles in this volume will obtain similar benefits from the application of the specialized techniques presented, and furthermore, will find valuable new information on a range of membrane properties.

METABOLISM AND INTRACELLULAR DISTRIBUTION OF A FLUORESCENT ANALOGUE OF PHOSPHATIDIC ACID IN CULTURED FIBROBLASTS*

Richard E. Pagano

Department of Embryology
Carnegie Institution of Washington
Baltimore, Maryland 21210

INTRODUCTION

Lipid molecules are essential building blocks for virtually every membrane of the living cell. Despite their importance, little is known about the mechanism(s) by which different lipids are sorted, transported, and assembled into various intracellular membranes. In order to study these problems we have developed an approach employing fluorescent phospholipids,[1-6] which, in some cases, can be used as true analogues of their natural counterparts. This approach has many advantages over conventional methods of metabolic study, including the abilities to directly observe fluorescent lipid metabolites within the living cell by fluorescence microscopy, and to detect and analyze minute amounts of fluorescent material. This report highlights some of our recent studies[7] with 1-acyl-2-(*N*-4-nitrobenzo-2-oxa-1,3-diazole)-aminocaproyl phosphatidic acid (C_6-NBD-PA, FIGURE 1), a fluorescent analogue of phosphatidic acid, which is a key intermediate in glycerolipid biosynthesis.

INCUBATION OF C_6-NBD-PA WITH CULTURED FIBROBLASTS

We have used small unilamellar vesicles as a vector for introducing C_6-NBD-PA into cultured Chinese hamster fibroblasts. In a typical experiment, vesicles are first prepared from dioleoyl phosphatidylcholine, C_6-NBD-PA, and the non-exchangeable fluorescent lipid[2,3] *N*-(lissamine) rhodamine B sulfonyl dioleoyl phosphatidylethanolamine (*N*-Rh-PE) (77/20/3, mol %), and then incubated with cells either in monolayer cultures or in suspension for 60 min at 2°C. All incubations are carried out in a protein-free HEPES-buffered balanced salt solution at a total vesicle lipid concentration of 0.2 μmol/ml. Following this incubation, the cells are washed, and either examined by fluorescence microscopy, or the lipids are extracted and analyzed using conventional analytical procedures.

TABLE 1 presents some typical results on the uptake of C_6-NBD-PA and *N*-Rh-PE by cells under these conditions. As seen from these data, significant amounts of the NBD-lipid are transferred to the cells with relatively little uptake of *N*-Rh-PE. This results in a ratio of NBD to rhodamine fluorescence in the washed, vesicle-treated cells that is much greater than in the applied vesicle suspension. This ratio demonstrates that the uptake of C_6-NBD-PA is due to preferential exchange or lipid transfer,[2,6,7] and is not due to the association of intact vesicles with cells.[6,8,9] Uptake of intact vesicles would cause the ratio of NBD to rhodamine fluorescence in the cell extracts to be identical to that found in the starting vesicles. Similar results to those

* Supported in part by a grant from the Whitehall Foundation and U.S. Public Health Service Grant GM22942.

0077-8923/83/0414-0001$01.75/0 © 1983, NYAS

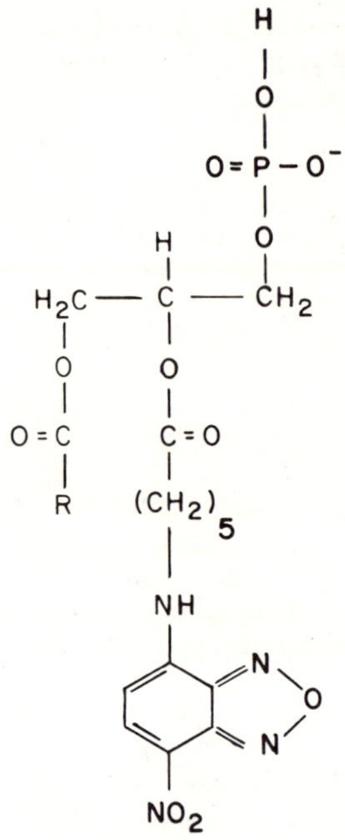

FIGURE 1. Structure of a fluorescent analogue of phosphatidic acid, 1-acyl-2-(*N*-nitrobenzo-2-oxa-1,3-diazole)-aminocaproyl phosphatidic acid. R, fatty acyl residue.

presented in TABLE 1 were obtained using monolayer cultures in place of cells in suspension (unpublished observations).

Extraction and analysis of the fluorescent lipid associated with the C_6-NBD-PA–treated cells revealed that 80–90% of the C_6-NBD-PA is hydrolyzed to 1-acyl-2-(*N*-4-nitrobenzo-2-oxa-1,3-diazole)-aminoacyl diglyceride (NBD-DG), even though the cells are maintained at 2°C throughout the experiment. Most of the remaining fluorescent lipid is in the form of C_6-NBD-PA, although small amounts of NBD-labeled phosphatidylcholine (NBD-PC) are sometimes also detected. If vesicle-cell incubations are carried out at 37°C, significantly more NBD-PC is formed, as well as 1,3-acyl-2-(*N*-4-nitrobenzo-2-oxa-1,3-diazole)-aminoacyltriglyceride (NBD-TG). Thus, C_6-NBD-PA is metabolized to the fluorescent products expected from the estab-

TABLE 1

CELLULAR UPTAKE OF C_6-NBD-PA BY CHINESE HAMSTER FIBROBLASTS AT 2°C*

	Cells	Applied Vesicles
pmol NBD-lipid/10^7 cells	3383	—
pmol *N*-Rh-PE/10^7 cells	16	—
(pmol NBD-lipid/pmol *N*-Rh-PE)	211	12.5

* Cells were incubated with small unilamellar vesicles composed of dioleoyl phosphatidylcholine/*N*-Rh-PE/C_6-NBD-PA (75.7/1.8/22.5, mol %) for 60 min at 2°C. The cells were then washed, the lipids extracted, and the amount of cell-associated NBD-lipid and *N*-Rh-PE quantified. Identical extractions and analyses were also made using an aliquot of the starting vesicle suspension.

lished lipid biosynthetic pathways (reviewed in References 10 and 11) in mammalian cells. In contrast, when NBD-PC is used in place of C_6-NBD-PA in vesicle-cell incubations, substantial amounts of fluorescent lipid become cell-associated, but no metabolism of this fluorescent lipid can be detected.[1,7]

Intracellular Distribution of NBD-Fluorescence in C_6-NBD-PA-treated Cells[7]

When vesicles containing C_6-NBD-PC are incubated with Chinese hamster fibroblasts in suspension for 60 min at 2°C, the fluorescent lipid is incorporated almost exclusively into the plasma membrane.[1] This results in peripheral ring fluorescence as seen in Figure 2a, with little, if any, fluorescent lipid being observed inside the cell. By contrast, when identical incubations are carried out with C_6-NBD-PA, no labeling of the plasma membrane is seen. Rather, the cell-associated NBD-lipids appear to be totally intracellular (Figure 2b), with labeling of the nuclear membrane being particularly prominent.

The distribution of intracellular fluorescence observed following treatment with C_6-NBD-PA is better resolved using monolayer cultures as seen in Figure 3. Two prominent intracellular features are seen. First, a portion of the fluorescence is localized in a reticular network in the cytoplasm; and second, bright "dots" of fluorescence are distributed throughout the cytoplasm. These two regions of the cell have been identified as the endoplasmic reticulum and mitochondria, respectively. Identification of the reticular network was accomplished as follows. Cells were treated with C_6-NBD-PA for 60 min at 2°C, washed, and photographed. The cells were then fixed, permeabilized, and treated with a rhodamine-conjugated lectin, *Lens culinaris* agglutinin, which has been shown to stain the endoplasmic reticulum.[12] Subsequent photography of the same cell using optics appropriate for rhodamine fluorescence demonstrated that the reticular network stained during C_6-NBD-PA treatment and the endoplasmic reticulum stained by the lectin were the same cytoplasmic structure. Identification of the bright "dots" of fluorescence was accomplished in an analogous manner using rhodamine 3B, a cationic fluorescent probe that is specifically accumulated by the mitochondria of living cells.[13]

Thus, when cells are incubated with C_6-NBD-PA at 2°C, the mitochondria, endoplasmic reticulum, and nuclear membrane (an extension of the endoplasmic reticulum) are the principal intracellular sites of fluorescence localization.

Metabolism and Redistribution of Intracellular Fluorescence at 37°C

In preliminary experiments we have found that if cells are treated with C_6-NBD-PA for 60 min at 2°C, washed, and then warmed to 37°C in a simple, balanced salt solution, labeling of the endoplasmic reticulum was decreased, while new cytoplasmic regions became labeled. Particularly prominent were fluorescent structures about 0.5–2 μm in diameter (Figure 4), which we have tentatively identified as intracellular lipid droplets. Specific events in NBD-lipid metabolism accompanied this redistribution of intracellular fluorescence. Namely, the cell-associated NBD-DG was metabolized to substantial amounts of NBD-labeled phosphatidylcholine and triglyceride. In the future we hope to determine which of these lipids is present in the lipid droplets, and whether, after long-term incubations at 37°C, some of the intracellular fluorescent lipids are transported to the cell surface.

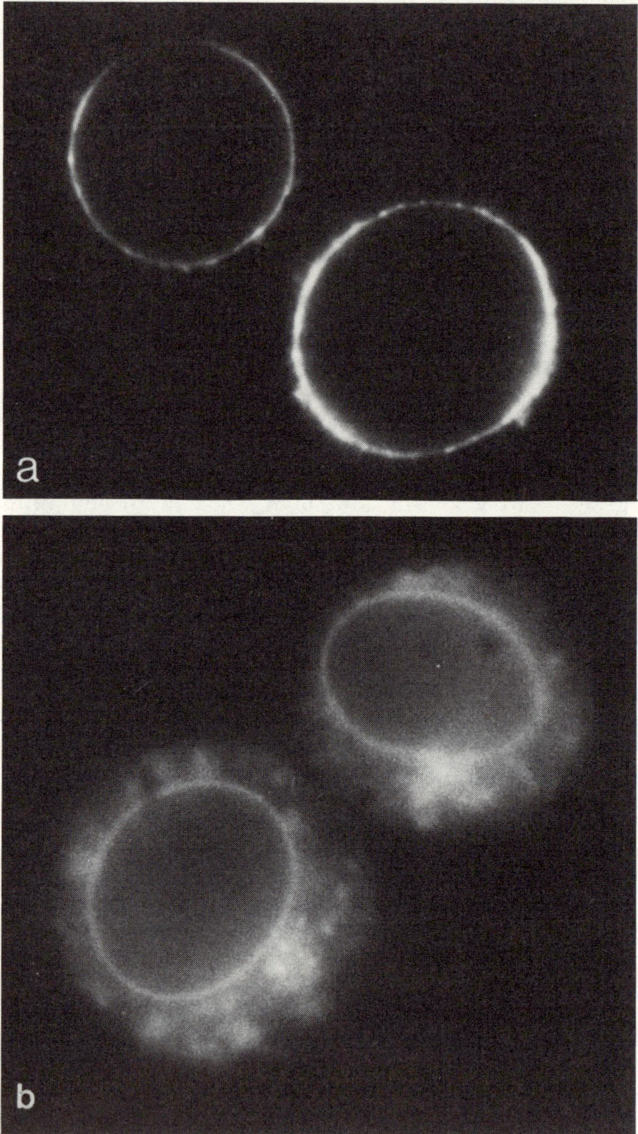

FIGURE 2. Fluorescence micrographs of Chinese hamster fibroblasts in suspension after incubation with C_6-NBD-PC– or C_6-NBD-PA–containing vesicles at 2°C. Incubations were performed at 2°C for 1 hr with dioleoyl phosphatidylcholine vesicles containing 5 mol % (a) C_6-NBD-PC or (b) C_6-NBD-PA.

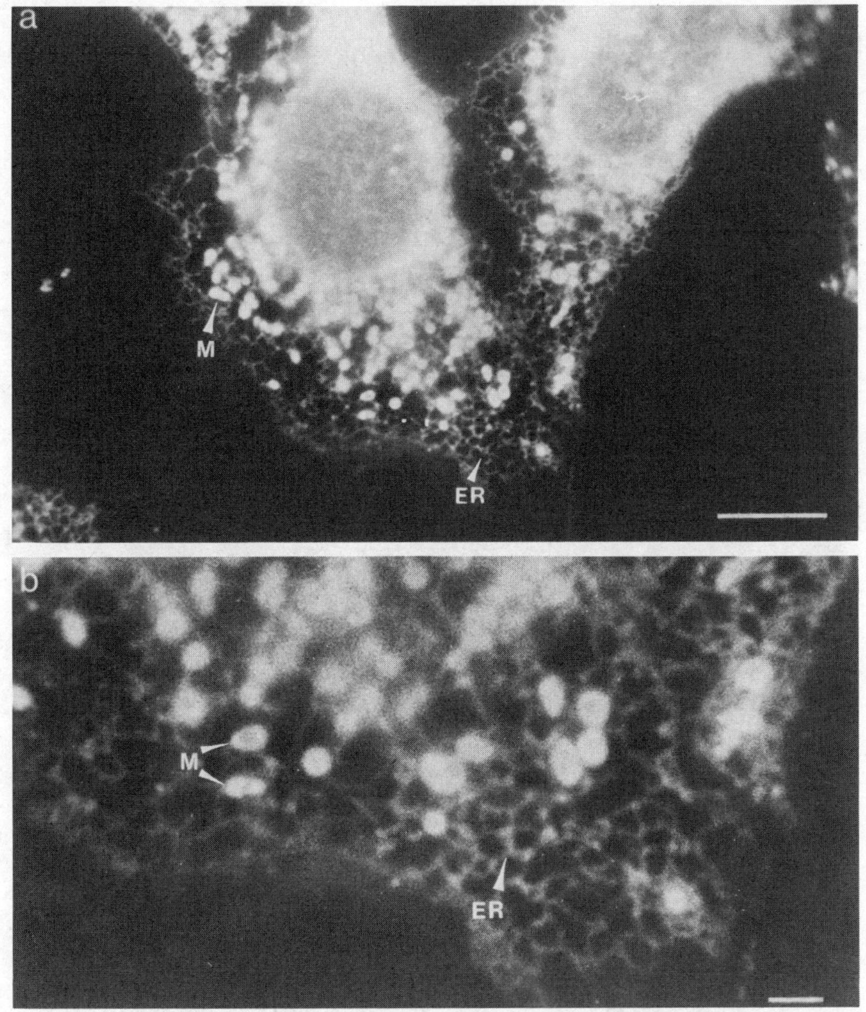

FIGURE 3. Fluorescence micrographs of Chinese hamster fibroblast monolayer cultures incubated with C_6-NBD-PA–containing vesicles for 60 min at 2°C. (a) Bar, 10 μm. (b) The same cell at higher magnification. Bar, 2 μm. ER and M designate areas of fluorescently stained endoplasmic reticulum and mitochondria, respectively. (From Pagano *et al.*[7] By copyright permission of The Rockefeller University Press.)

SUMMARY

We have shown that a fluorescent compound, C_6-NBD-PA, behaves as an analogue for phosphatidic acid, an important intermediate in glycerolipid biosynthesis. This derivative is preferentially transferred from phospholipid vesicles to cultured Chinese hamster fibroblasts at 2°C, while the C_6-NBD-PA–derived fluorescence is localized at the nuclear membrane, endoplasmic reticulum, and mitochondria. Ex-

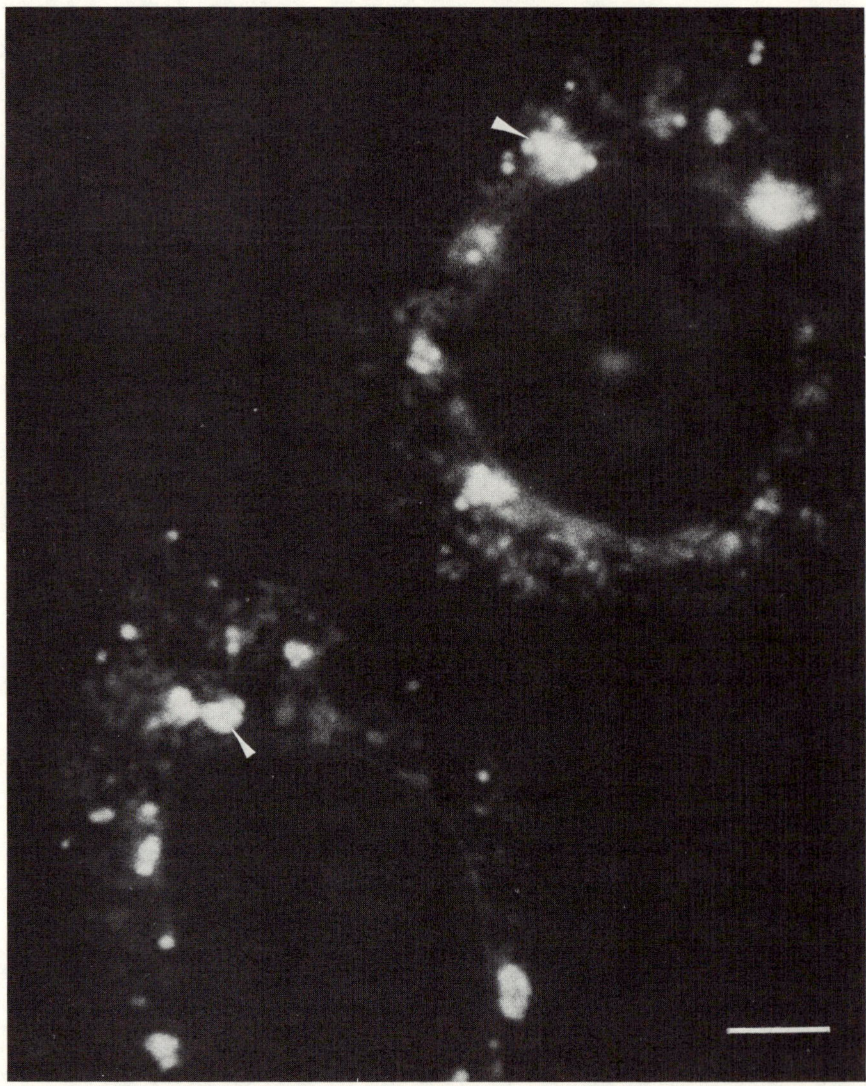

FIGURE 4. Fluorescence micrograph of Chinese hamster fibroblast monolayer culture incubated with C_6-NBD-PA for 60 min at 2°C, washed, and subsequently warmed to 37°C for 30 min. Arrows indicate fluorescent intracellular lipid droplets. Bar is 10 μm.

traction and analysis of the fluorescent lipids associated with the cells after treatment with vesicles at 2°C or 37°C revealed that a large fraction of the fluorescent phosphatidic acid is converted to fluorescent diglyceride, phosphatidylcholine, and triglyceride. Although we do not yet know how accurately the metabolism and intracellular distribution of C_6-NBD-PA and its metabolites reflect those of endogenous phosphatidic acid, it is encouraging that this fluorescent analogue is apparently metabolized through the diglyceride pathway to give fluorescent analogues of diglyceride, triglyceride, and phosphatidylcholine. This metabolism suggests that the presence of the fluorescent group on the acyl chain of the phosphatidic acid analogue does not inhibit the enzymes involved in phosphatidic acid metabolism.

We conclude that fluorescent lipid analogues such as C_6-NBD-PA may be useful in correlating biochemical studies of lipid metabolism with studies of the intracellular localization of lipid metabolites by fluorescence microscopy.

References

1. Struck, D. K. & R. E. Pagano. 1980. Insertion of fluorescent phospholipids into the plasma membrane of a mammalian cell. J. Biol. Chem. **256**: 5404–5410.
2. Struck, D. K., D. Hoekstra & R. E. Pagano. 1981. Use of resonance energy transfer to monitor membrane fusion. Biochemistry **20**: 4093–4099.
3. Pagano, R. E., O. C. Martin, A. J. Schroit & D. K. Struck. 1981. Formation of asymmetric phospholipid membranes via spontaneous transfer of fluorescent lipid analogues between vesicle populations. Biochemistry **20**: 4920–4927.
4. Nichols, J. W. & R. E. Pagano. 1981. Kinetics of soluble lipid monomer diffusion between vesicles. Biochemistry **20**: 2783–2789.
5. Nichols, J. W. & R. E. Pagano. 1982. Use of resonance energy transfer to study the kinetics of amphiphile transfer between vesicles. Biochemistry **21**: 1720–1726.
6. Pagano, R. E., A. J. Schroit & D. K. Struck. 1981. Interactions of phospholipid vesicles with mammalian cells *in vitro*: Studies of mechanism. *In* Liposomes: From Physical Structure to Therapeutic Applications. C. G. Knight, Ed. Chapter 11. Elsevier/North-Holland Biomedical Press. Amsterdam.
7. Pagano, R. E., K. J. Longmuir, O. C. Martin & D. K. Struck. 1981. Metabolism and intracellular localization of a fluorescently labeled intermediate in lipid biosynthesis within cultured fibroblasts. J. Cell Biol. **91**: 872–877.
8. Szoka, F., K. Jacobson, Z. Derzko & D. Papahadjopoulos. 1980. Fluorescence studies on the mechanism of liposome-cell interactions in vitro. Biochim. Biophys. Acta **600**: 1–18.
9. Schroit, A. J. & R. E. Pagano. 1981. Capping of a phospholipid analog in the plasma membrane of lymphocytes. Cell **23**: 105–112.
10. Bell, R. M. & R. A. Coleman. 1980. Enzymes of glycerolipid synthesis in eukaryotes. Ann. Rev. Biochem. **49**: 459–487.
11. van Golde, L. M. G. & S. G. van den Bergh. 1977. Introduction: General pathways in the metabolism of lipids in mammalian tissues. *In* Lipid Metabolism in Mammals. F. Snyder, Ed. **1**: 1–33. Plenum Press. New York.
12. Virtanen, I., P. Ekblom & P. Laurila. 1980. Subcellular compartmentalization of saccharide moieties in cultured normal and malignant cells. J. Cell Biol. **85**: 429–434.
13. Johnson, L. V., M. L. Walsh, B. J. Bockus & L. B. Chen. 1981. Monitoring of relative mitochondrial membrane potential in living cells by fluorescence microscopy. J. Cell Biol. **88**: 526–535.

HETEROGENEITY IN THE PLASMA MEMBRANE LIPIDS OF EUKARYOTIC CELLS

Michael Edidin and Alice VanVoris Sessions

Biology Department
The Johns Hopkins University
Baltimore, Maryland 21218

The so-called fluid mosaic model[1] of cell membrane organization has served well as the intellectual basis for much work on membrane organization and function. This model emphasized the hydrophobic interaction of membrane integral proteins with the lipid bilayer, and, though other alternatives were considered, the independence of membrane components and the tendency for these components, both lipids and proteins, to randomize by diffusion in the plane of the membrane. The idea of a fluid membrane whose components were free to diffuse in the plane of the membrane developed from work on the movement of membrane proteins in the plasma membranes of isolated mammalian cells, lymphocytes,[2] or cultured fibroblasts.[3] The impact of these experimental demonstrations of capping and diffusion of membrane proteins obliterated the observations on the restrictions to lateral diffusion that were evident in tissue cells, for example in the localization of membrane enzymes to one face only of an epithelial cell (discussed by Edidin[4]) or in the isolation of membrane fractions of plasma membrane with greatly differing lipid and protein compositions.[5,6]

Renewed interest in the heterogeneity of membrane organization was awakened by a series of experiments and speculations on the organization of membrane lipids. A series of papers on synthetic lipid vesicles by McConnell and co-workers developed phase diagrams for mixtures of lipids and showed clearly that several lipid phases could co-exist in a single vesicle.[7,8] Jain and White[9] offered a model in which membrane lipids are segregated by species into a series of immiscible domains.

The existence of lipid domains was also postulated in order to explain discrepancies in the behavior of lectin receptors or enzymes in cells with modified fatty acid composition. Thus, Horowitz and co-workers[10] reported that the lectin-mediated agglutination of 3T3 or SV3T3 mouse fibroblasts was a function of temperature and membrane fatty acyl composition, but that the critical temperature for agglutination was different for the lectins concanavalin A and wheat germ agglutinin. They argued that the receptors for the two lectins were sited in different lipid environments. This argument was supported by later work[11] in which a spin label reported two apparent transitions in the membrane lipid phases whose temperatures correspond to the critical temperatures for agglutination by the lectins. Similar arguments have been applied to differential effects of temperature or lipid substitution enzyme activity (reviewed by Jain and Wagner[12]). In other cases, we infer that physical probes of membrane lipids, for example spin labels, are in different environments than membrane enzymes or receptors whose activity is being followed biochemically.[13] All of these experiments suggest, if only indirectly, that membrane lipids are heterogeneously distributed. They do not address the extent of domains around particular proteins, though recent work on boundary lipids[14] and on the functional requirements for reconstitution of lipid-requiring membrane enzymes (review by Jain and Wagner[12]) suggest that boundary domains are short lived and rather disordered and that there is usually little lipid specificity shown in reconstituted enzyme preparations.

0077-8923/83/0414-0008$01.75/0 © 1983, NYAS

FIGURE 1. 'diI' a carbocyanine dye family with alkyl chains varying from 10 to 22 carbons in length.

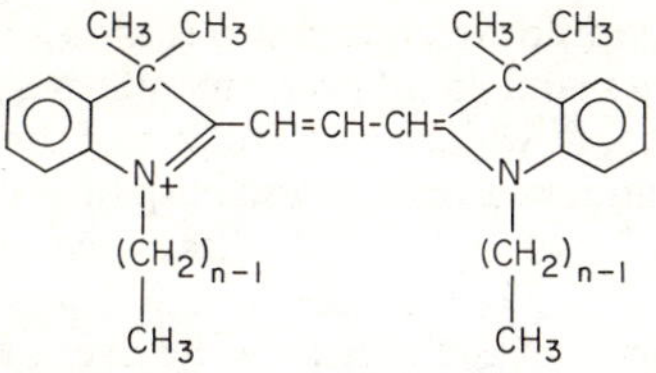

CH_3 CH_3 C CH=CH-CH= N+ $(CH_2)_{n-1}$ CH_3 — CH_3 CH_3 C N $(CH_2)_{n-1}$ CH_3

3,3′-diacylindocarbocyanine Iodide

Recent work with lipid-soluble fluorescent probes has given somewhat more direct evidence for domains in the lipids of native membranes. The data show that a given probe or members of a series of related probes reside in different environments in a single membrane, though we still cannot determine either the size or the lifetime of these domains. The two groups of published experiments strongly suggesting the presence of lipid domains are by Klausner and co-workers[15,16] and by Wolf and co-workers.[17,18,19] Klausner *et al.*[15] examined the fluorescence polarization and fluorescence lifetime of diphenylhexatriene (DPH), a hydrophobic fluorescent lipid probe, in synthetic lipid vesicles and in native lymphocyte membranes. The work with synthetic vesicles showed that membrane perturbants that had a differential effect on DPH polarization in native membranes did not show any differential in single component vesicles. These compounds did have differential effect on DPH in mixed vesicles containing both fluid and gel phases. This argued that the differential effects (of *cis* and *trans* unsaturated fatty acids) reflected the presence of fluid and gel domains in native plasma membranes. This interpretation was reinforced by the analysis of fluorescence lifetime data. DPH fluorescence in native membranes and in mixed-phase vesicles could only be analyzed in terms of two or more lifetimes while DPH in single-component vesicles gave a single lifetime by phase fluorometry.

Wolf and co-workers used a similar logic to infer the presence of multiple domains in native membranes of sea urchin (*S. puperatus*)[17] and mouse (*M. musculus*)[18] eggs. They measured the lateral diffusion of a series of lipid probes, the diI's (3,3′-

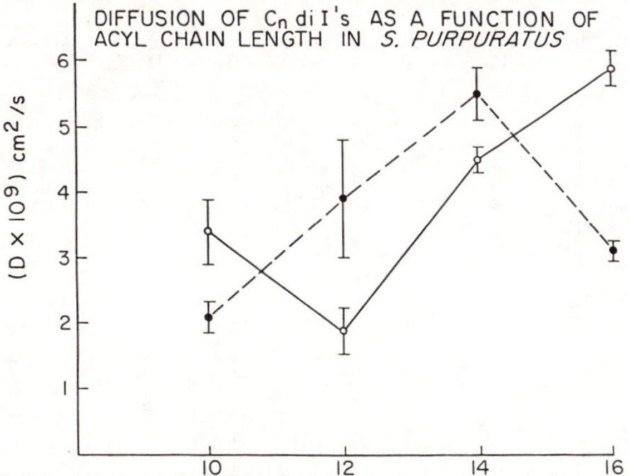

FIGURE 2. Lateral diffusion of diI's of various alkyl chain lengths in membranes of unfertilized and fertilized *S. purperatus* eggs.[17] ———, unfertilized. – – –, fertilized. Diffusion was measured by the method of fluorescence photobleaching and recovery (FPR).[26,27]

dialkylindocarbocyanines) (FIGURE 1) in single component and multicomponent lipid vesicles and in plasma membranes of intact cells. The results from native membranes, in which the diffusion of the diI's was a function of alkyl chain length, could be mimicked only in mixed-phase vesicles, not in single component vesicles[19] (FIGURES 2 and 3). The diI's, like the probes used by Klausner *et al.*, appear to partition preferentially into gel or fluid phases as shown by fluorimetric[19] and calorimetric[20] methods. It is not clear if the dyes probe gel and fluid domains in native membranes, or if they partition into immiscible fluid domains of different compositions. As noted above, the size of the putative domains cannot be estimated from these experiments. The diffusion behavior of the probe reports on domain viscosity, on partition between different domains and on the lifetime of these domains. If the partition coefficients can be established for the dyes, we should be able to determine the size of the domains, assuming that a given domain persists for a period on the order of a minute or more.

At this point, we have strong indications from work on lipid probes of the differentiation of plasma membranes into lipid domains, and we also have considerable evidence from studies of protein distribution and function that particular proteins may be associated with such domains. In some cases mentioned above, changes in lipid composition or organization are not readily associated with changes in protein

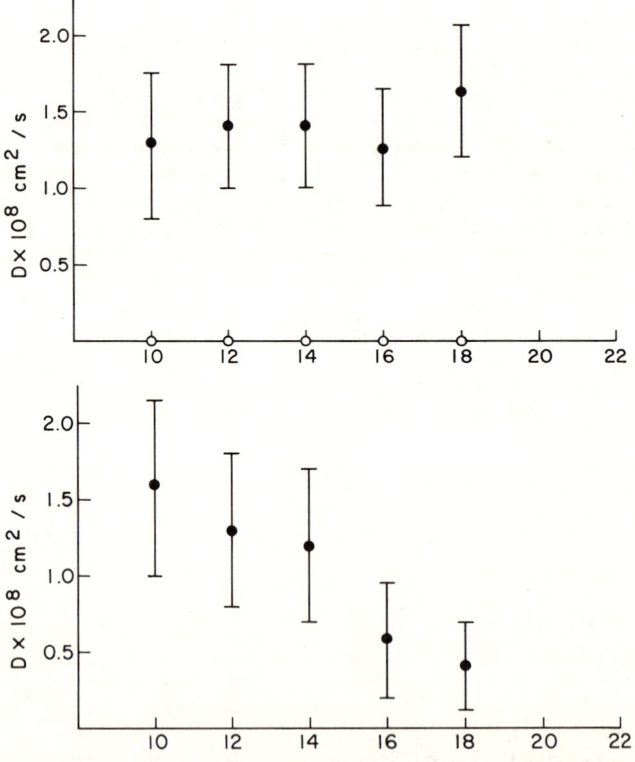

FIGURE 3. Lateral diffusion of diI's of various alkyl chain lengths in synthetic lipid vesicles.[19] (Top) Single component (●) fluid or (○) gel vesicles. (Bottom) Mixed gel and fluid vesicles.

function, while in other instances transitions in membrane lipids reported by fluorescence probes are reflected in the behavior of membrane enzymes.[21]

In the rest of this paper we will summarize some work on activity of membrane enzymes, membrane lipid composition, and membrane physical properties obtained in a single-cell system, cultured fish fibroblasts. The data reinforce the observations briefly summarized above in that they indicate that membrane enzymes are differentially affected by alterations in membrane acyl chain and cholesterol content. They also show that, even in a membrane in which these differences suggest organization into domains, both spin labels and probes of protein lateral diffusion fail to indicate the presence of such domains.

Fathead minnow (*Pimephales promelas*) fibroblasts (ATCC-42) were of interest to us since they are derived from a poikilothermic animal whose environment (freshwater ponds) undergoes considerable temperature changes during the course of a day and during the course of the seasons. Consonant with this temperature range in the native habitat, it has been found that the cells will grow in culture at temperatures ranging from under 10°C to 37°C. We were able to confirm this wide range of growth temperatures, but found that 15°C was about the lowest practical limit for

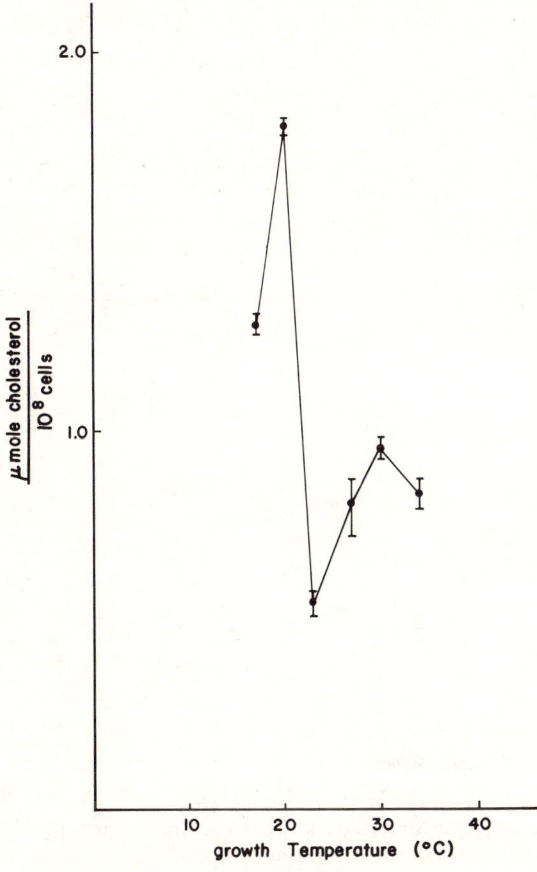

FIGURE 4. Cholesterol content of FHM cells cultured for 3 days at the temperatures indicated. Cholesterol was assayed by the cholesterol oxidase method of Allain and coworkers.[28] Bars indicate standard deviations of replicate assays.

cell growth, especially since we needed substantial numbers of cells for membrane fractionation and biochemical study. We expected that the cell lipid composition would be affected by growth temperature and that cells grown at lower temperatures would contain larger proportions of unsaturated acyl chains than cells grown at higher temperatures.[22] In fact, we found that the temperature of cell culture had the greatest effect on cholesterol content and that changes in acyl chain composition were in some instances the opposite of what we would have predicted from other work on the temperature adaptation of poikilothermic animals. (These effects were dependent upon the lot of fetal calf used in the cultures and may reflect selection of variants during the many years this cell line has been in culture.) The changes in cholesterol content as a function of growth temperature are shown for intact cells in FIGURE 4 and for plasma membranes, isolated by the method of Schimmel *et al.*,[28] in FIGURE 5. It will be seen that significant increases in cholesterol content occurred in cells grown at 20 and 30°C. The ratio of saturated to unsaturated acyl chains in plasma membrane fractions increased considerably in cells grown at 20°C, where around 75% of all acyl chains were saturated. The ratio of saturated to unsaturated

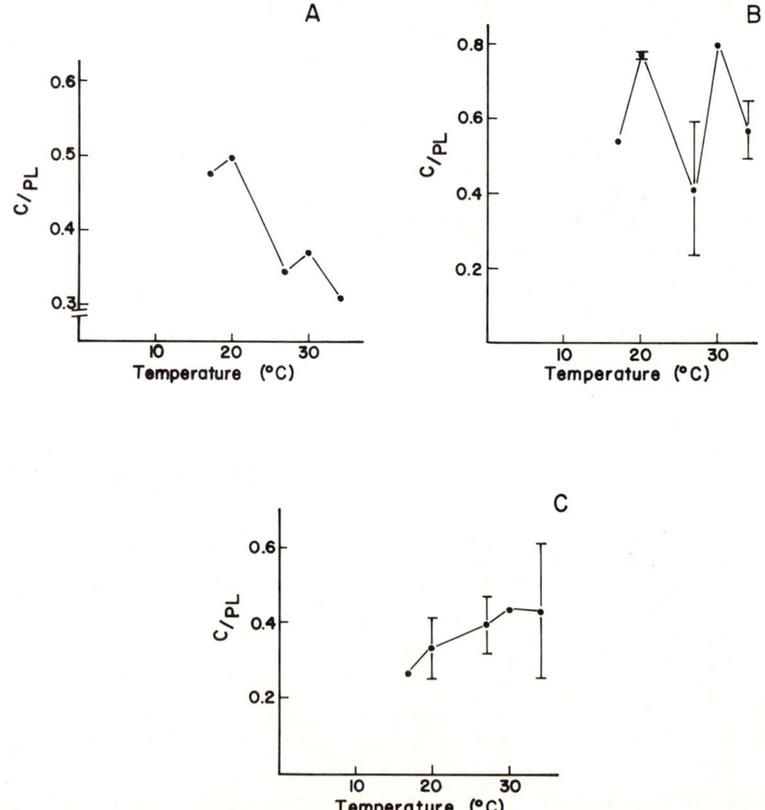

FIGURE 5. Cholesterol/phospholipid ratios (C/PL) in membrane fractions isolated from FHM cells. (A) C/PL for all five membrane fractions. (B) C/PL for the plasma membrane-rich fraction. (C) C/PL for the endoplasmic reticulum-rich fraction.

fatty acids was constant at around 55:45 in cells grown at all other temperatures in a range from 17 to 34°C (FIGURE 6). The increase in saturated fatty acids at 20°C was largely due to an increase in palmitic and stearic acid content. The high level of these fatty acids in membranes implies that a significant proportion of phospholipids in these membranes are disaturated, rather than the usual 1-saturated, 2-unsaturated phospholipids. In the absence of cholesterol these species would form discrete gel domains in a native membrane. The increased cholesterol found in membranes of cells grown at 20°C would disrupt such gel domains.

The changes described were not found in other membrane fractions isolated from the cells. They were also unique to cells grown at 20°C. Even though cells grown at 30°C also had higher cholesterol/phospholipid ratios than cells grown at temperatures other than 20°C, they did not contain increased levels of saturated fatty acids and indeed appeared to contain considerably higher levels of palmitoleic acid (16:1) than cells grown at other temperatures. Thus the two peaks of increased cholesterol content as a function of temperature are not associated with similar changes in acyl chain composition.

The compositional data for FHM cells grown at different temperatures suggest that membrane lipids are organized in domains. We attempted to probe these domains by examining the behavior of a fatty acid spin label 3-doxyl palmitic acid as a function of temperature in cells grown at different temperatures. The order parameter was an unbroken monotonic function of measurement temperature for all cells examined. However, the sensitivity of the order parameter to temperature change was least in cells grown at 20° and 30°C, presumably reflecting the averaging effect of cholesterol on order in membrane lipids (FIGURE 7). The order parameter reported for cells at the temperature of their growth is shown in FIGURE 8. Contrary to some predictions it is not a constant, but decreases with increasing temperature. Thus if membrane enzymes require a constant or uniform lipid viscosity for function, at least some of these enzymes must be segregated in regions not probed by the spin label, or regions that are small, making little contribution to the signal, compared with the whole membrane.

If membrane enzymes are affected by their lipid environment then, to the extent that they are about equally embedded in the bilayer, we expect their function to be

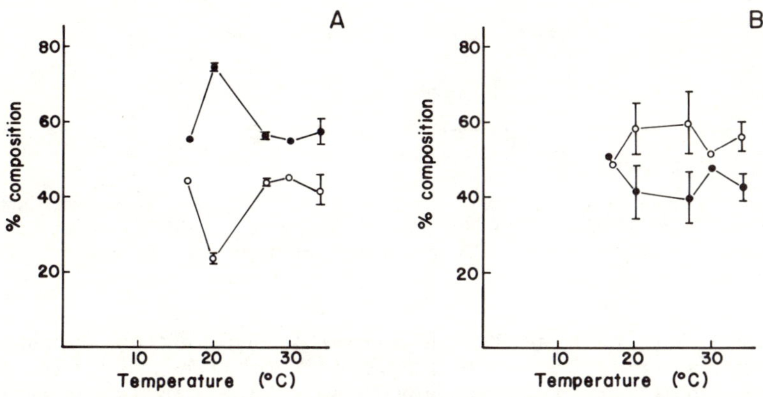

FIGURE 6. Proportions of (●——●) saturated and (○——○) unsaturated fatty acids in extracts of: (A) plasma membrane-rich and (B) endoplasmic reticulum-rich membranes (fractions 2 and 4 of Schimmel *et al.*[28]).

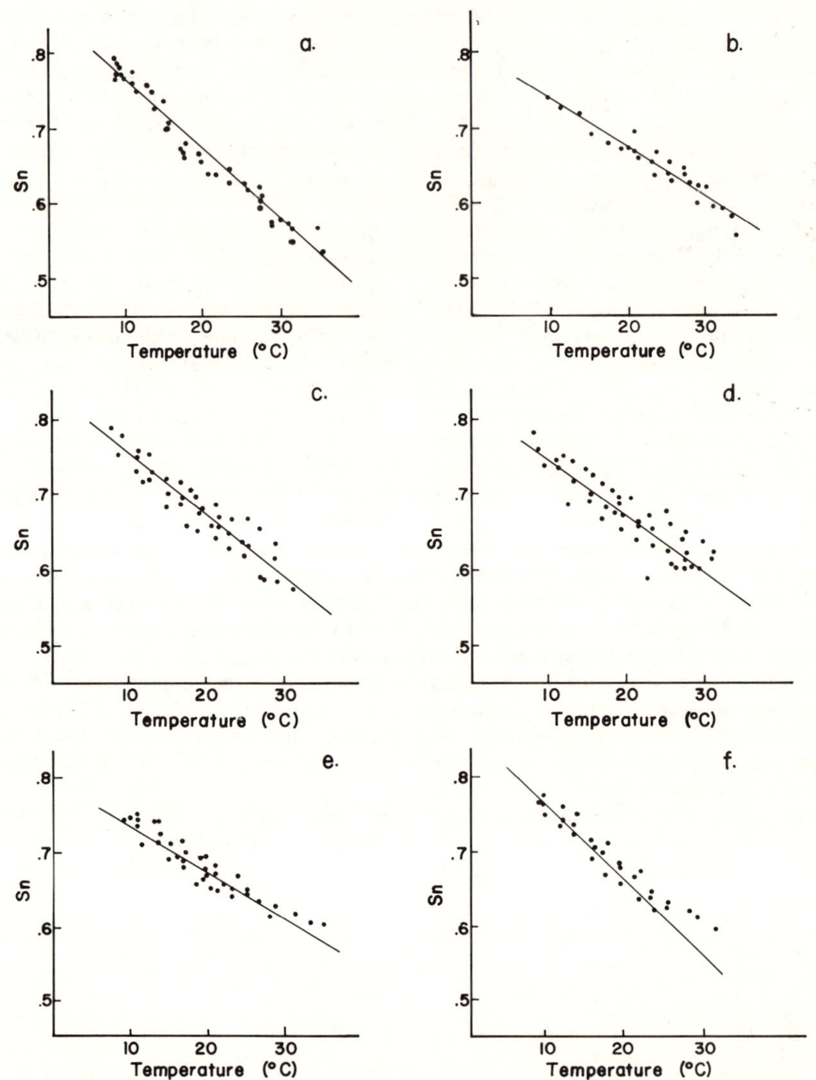

FIGURE 7. Sensitivity to temperature of order parameter (S_n) reported by 3-doxylpalmitic acid in cells grown at: (a) 34°C, (b) 30°C, (c) 27°C, (d) 23°C, (e) 20°C, (f) 17°C. The slope of the regression line for cells grown at 20°C is significantly less than the slopes for all other lines except for cells grown at 30°C.

affected by changes in membrane lipid composition. Accordingly we examined the activities of a number of enzymes of endoplasmic reticulum and plasma membranes. There was no striking variation with growth temperature in activity of cytochrome *c* reductase or glucose-6-phosphate dehydrogenase (assayed at 37°C). In contrast, plasma membrane enzymes assayed at 37°C varied significantly as a function of cell growth temperature. FIGURE 9 shows the variation in activity of 5′-nucleotidase and

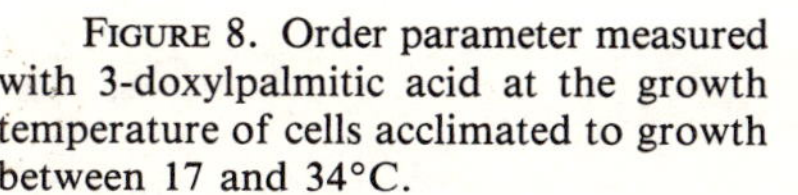

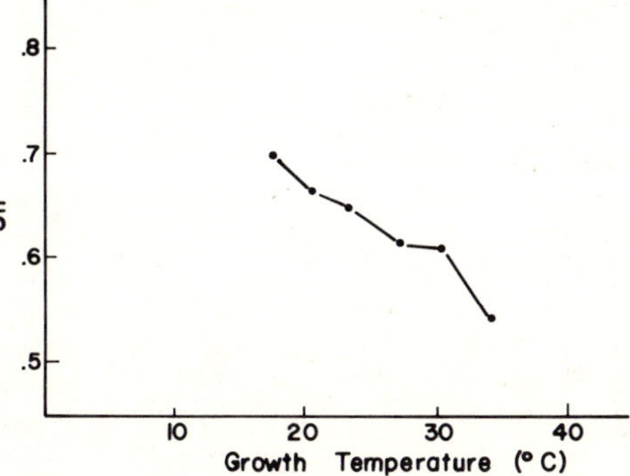

FIGURE 8. Order parameter measured with 3-doxylpalmitic acid at the growth temperature of cells acclimated to growth between 17 and 34°C.

FIGURE 10 that of leucine aminopeptidase (LAP) as a function of temperature of cell growth. The peaks of LAP activity correspond to the peaks in membrane cholesterol content, while the 5′-nucleotidase activity is a maximum in cells grown at 30°C, but is not elevated in membranes from cells grown at 20°C. These data suggest that the two enzymes are in different lipid environments, one of which is affected by changes in membrane cholesterol content and the other of which is not. This indication was reinforced by an experiment in which cells were grown at 20°C in the presence of an inhibitor of cholesterol synthesis, 25-OH cholesterol. Activity was not increased in cells grown at 20°C in the presence of 25-OH cholesterol, but it was increased in cells grown at 20°C in the absence of 25-OH cholesterol.

The compositional data and the comparison of the activities of two different membrane integral enzymes suggest that different protein components of the plasma membrane are sited in lipid domains of different composition. We were interested in examining the effects of lipid compositional changes on diffusion of membrane components and accordingly we measured lateral diffusion of membrane proteins labeled with fluorescent Fab fragments of a rabbit anti-FHM antibody. Diffusion of the labeled molecules as a function of temperature of cell growth and of measurement is shown in FIGURE 11. It appears that the changes in membrane composition

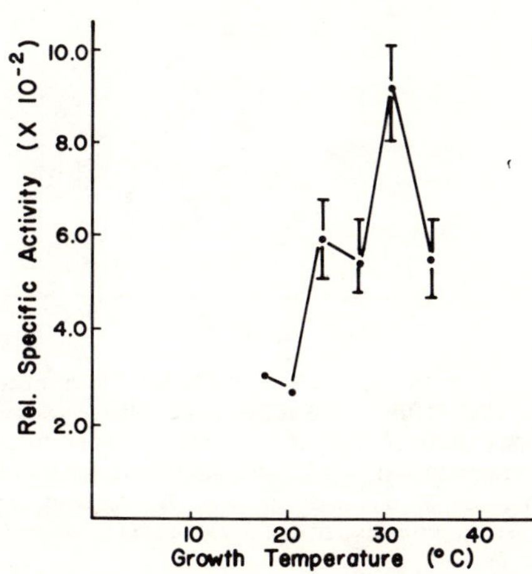

FIGURE 9. Specific activity of 5′-nucleotidase in cells acclimated to growth at various temperatures. Enzyme activity was measured by the method of Dixon and Purdom.[30]

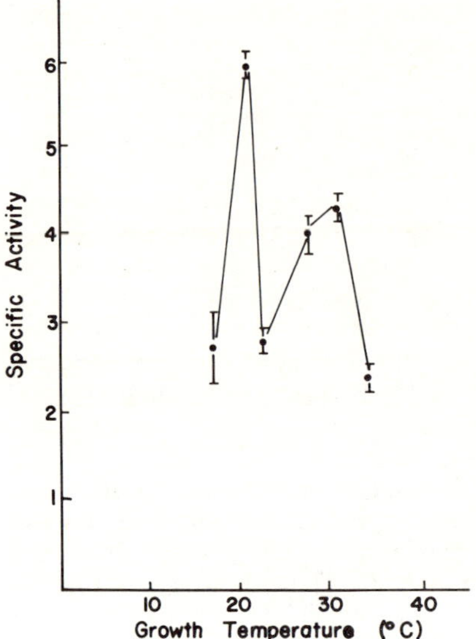

FIGURE 10. Specific activity of leucine aminopeptidase in cells acclimated to growth at various temperatures. The enzyme was assayed by the method of Berger and Broida.[31]

with changes in cell growth temperature do not affect either the rate or the temperature dependence of lateral diffusion of a selection of membrane proteins. This result is similar to that of Axelrod *et al.*[23] who found that changes in membrane acyl chain composition had little effect on lateral diffusion of membrane proteins. It appears that whatever local heterogeneity there may be in lipid organization, it is not effectively sensed by proteins diffusing over a distance of a few micrometers. This may be because constraints to lateral diffusion are dominated by interaction of integral proteins with a cytoskeleton. However, the diffusion coefficients that we obtained for FHM proteins are about the same as those found for vertebrate rhodopsin.[24,25] It is about as high as expected for diffusion constrained only by lipid viscosity. Axelrod

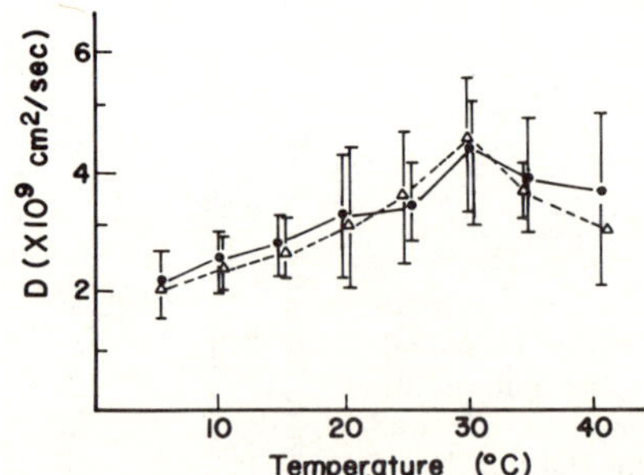

FIGURE 11. Lateral diffusion of membrane proteins labeled with Fab rabbit anti-FHM. Lateral diffusion was measured by fluorescence photobleaching and recovery (FPR).[27] The fractional recoveries for all cells at all temperatures were 40–50%. Lateral diffusion was inhibited by fixation with 5% paraformaldehyde. Diffusion coefficients were not affected by the presence of an excess of unlabeled antibody. Hence, binding of Fab fragments to the membranes is irreversible over the time course of the experiment. (●——●) cells grown at 34°C; (△- - -△) cells grown at 20°C.

and co-workers also found that there was little or no effect of changing membrane fatty acid composition on diffusion of a lipid-soluble probe.[23] Thus the most likely interpretation of our results is that the diffusing species do not probe regions whose lipid composition is altered during cell growth. We have not yet studied FHM cell membranes with the lipid probes that report a variety of environments in egg plasma membranes.

In summary, our data on FHM cells are concordant with a view developing from measurements of the behavior of lipid probes in a number of plasma membranes. It appears that different components of the plasma membrane reside in different lipid environments and that these differing environments can be detected in terms of differentials in the behavior of membrane enzymes as well as in differentials in the lateral diffusion of selected membrane probes. The evidence for membrane lipid heterogeneity continues to increase. A useful membrane model incorporating this heterogeneity has yet to be developed.

References

1. Singer, S. J. & G. Nicolson. 1972. Science **175**: 720–731.
2. Taylor, R. B., W. P. H. Duffus, M. C. Ruff & S. dePetris. 1971. Nature (New Biol.) **223**: 225–229.
3. Frye, L. D. & M. Edidin. 1970. J. Cell Sci. **7**: 319–335.
4. Edidin, M. 1972. *In* Membrane Research. C. F. Fox, Ed.: 15–25. Academic Press. N.Y.
5. Kawai, K., M. Fujita & M. Nakao. 1974. Biochim. Biophys. Acta **369**: 222–233.
6. Evans, W. H. 1980. Biochim. Biophys. Acta **604**: 27–64.
7. Shimschick, E. J. & H. M. McConnell. 1973. Biochemistry **12**: 2351–2360.
8. Wu, S. H. W. & H. M. McConnell. 1979. Biochemistry **14**: 847–854.
9. Jain, M. K. & H. B. White. 1977. Adv. Lipid Res. **15**: 1–59.
10. Horwitz, A. F., M. E. Hatten & M. M. Burger. 1974. Proc. Natl. Acad. Sci. USA **71**: 3115–3119.
11. Hatten, M. E., C. J. Scandella, A. F. Horwitz & M. M. Burger. 1978. J. Biol. Chem. **253**: 1972–1977.
12. Jain, M. K. & R. C. Wagner. 1980. Introduction to Biological Membranes. John Wiley and Sons. New York.
13. Bakardjieva, A., H. J. Galla & E. J. M. Helmreich. 1979. Biochemistry **18**: 3016–3023.
14. Chapman, D. 1982. *In* Membranes and Transport. A. N. Martonosi, Ed. **1**: 63–70. Plenum. New York.
15. Klausner, R. D., A. M. Kleinfeld, R. L. Hoover & M. J. Karnovsky. 1980. J. Biol. Chem. **255**: 1285–1295.
16. Karnovsky, M., A. M. Kleinfeld, R. L. Hoover & R. D. Klausner. 1982. J. Cell Biol. **94**: 1–6.
17. Wolf, D. E., W. Kinsey, W. Lennarz & M. Edidin. 1981. Dev. Biol. **81**: 133–138.
18. Wolf, D. E., M. Edidin & A. H. Handyside. 1981. Dev. Biol. **85**: 195–198.
19. Klausner, R. D. & D. E. Wolf. 1980. Biochimie **19**: 6199–6203.
20. Ethier, M. F., D. E. Wolf & D. L. Melchior. 1983. Biochemistry **22**: 1178–1182.
21. Brasitus, T. A. & D. Schachter. 1980. Biochemistry **19**: 2763–2765.
22. Cossins, A. R. 1981. *In* Effects of Low Temperature on Biological Membranes. A. Clarke & J. Morris, Eds.: 83–106. Academic Press. London.
23. Axelrod, D., A. Wight, W. Webb & A. Horwitz. 1978. Biochemistry **17**: 3604–3609.
24. Poo, M.-M. & R. A. Cone. 1974. Nature (New Biol.) **247**: 438–441.
25. Wey, C. L., R. A. Cone & M. A. Edidin. 1981. Biophys. J. **33**: 225–232.
26. Peters, R. 1981. Cell Biol. Intl. Rep. **5**: 733–760.
27. Wolf, D. E. & M. Edidin. 1981. *In* Techniques in Cellular Physiology. P. Baker, Ed. **1**: 1–4. Elsevier/North Holland. Amsterdam.
28. Schimmel, S. D., C. Kent, R. Bischoff & P. R. Vagelos. 1973. Proc. Natl. Acad. Sci. USA **70**: 3195–3199.

29. Allain, C. C., L. S. Poon, C. S. Chan, W. Richmond & P. C. Fu. 1974. Clin. Chem. **20**: 470.
30. Dixon, T. F. & M. Purdom. 1954. J. Clin. Path. **7**: 341.
31. Berger, L. & D. Broida. 1975. Sigma Technical Bulletin **251**.

LIPID FLUIDITY OF THE INDIVIDUAL HEMILEAFLETS OF HUMAN ERYTHROCYTE MEMBRANES*

David Schachter, Richard E. Abbott, Uri Cogan,† and Michael Flamm

Department of Physiology
College of Physicians and Surgeons
Columbia University
New York, New York 10032

INTRODUCTION

Although it is increasingly well recognized that membrane lipid fluidity is an important determinant of membrane functions, a number of unresolved questions concerning lipid fluidity deserve serious consideration. To begin with, the very term "lipid fluidity" as applied to anisotropic bilayer membranes has been used in several senses by various authors. Many investigators have used it as a general term to express the relative motional freedom of the bilayer lipid molecules or substituents thereof, combining in the one term concepts both of rate of movement (a dynamic component) and extent of movement (a component related to lipid structure and expressed as an order parameter, which characterizes the degree to which movement is hindered). When assessed by steady-state fluorescence polarization of lipid fluorophores in bilayer membranes, this usage amounts to quantifying "lipid fluidity" via the fluorescence anisotropy, r, without further resolution of the components that determine r. Studies with 1,6-diphenyl-1,3,5-hexatriene (DPH), for example, have demonstrated that rotations of this rodlike fluorophore in natural and artificial bilayer membranes are hindered[1-4] and can be described by the relationship

$$r = r_\infty + (r_0 - r_\infty)\,[\tau_c/(\tau_c + \tau_f)],$$

where r_∞ is the limiting hindered anisotropy observed after relatively long time intervals in time-resolved anisotropy decay experiments, r_0 is the maximal limiting anisotropy, τ_c is the correlation time, and τ_f is the excited-state lifetime.[5,6] Heyn[5] and Jähnig[6] suggest that the term fluidity be applied only in reference to τ_c, i.e., to rates of rotation, whereas r_∞ is related to an order parameter, S, where $S = (r_\infty/r_0)^{1/2}$. Van Blitterswijk *et al.*,[7] on the other hand, propose that for DPH studies, at least, the parameter r_∞ is a useful index of "fluidity." For many biological applications changes in r owing to changes in τ_c, r_∞, or both are of significance, and a general term is needed for such alterations. It seems reasonable to use the term "lipid fluidity" to designate motional freedom in this broad sense and to characterize further the relative contributions of the dynamic component and order parameter whenever possible.

A second important issue in relation to membrane lipid fluidity is the heterogeneous lipid composition of most biological membranes. The human erythrocyte membrane has been particularly well studied in this respect and it is clear that the outer and inner leaflets differ in their composition.[8,9] In these cells, as well as in a number of other cell types, the outer leaflet is relatively rich in phosphatidylcholine and sphingomyelin, whereas the inner leaflet contains most of the phosphatidyletha-

* Supported by research grants HL16851 and AM21238, and training grant (M.F.) GM07367 of the National Institutes of Health.

† Present address: Israel Institute of Technology, Haifa, Israel.

0077-8923/83/0414-0019$01.75/0 © 1983, NYAS

nolamine and essentially all of the phosphatidylserine. Given this characteristic pattern of asymmetry in composition, a number of investigators have sought to establish whether it implies a corresponding asymmetry of fluidity. The initial results of such studies did provide evidence for asymmetry of fluidity, but the results were not entirely consonant. Rottem[10] studied electron-spin resonance (ESR) spectra of spin-labeled stearic acid probes in *Mycoplasma hominis* and *Acholeplasma laidlawii* and concluded that the outer leaflet is more fluid in each species. Wisnieski and Iwata[11] compared a nitroxide derivative of decane, which is capable of "flip-flop" across the bilayer, with a glucosamine derivative of stearic acid 12-nitroxide, which is restricted in its ability to flip. Studying plasma membrane preparations derived from mouse LM cells and embryonated chick cells (Newcastle disease virus envelopes), they too concluded that "the outer membrane monolayer is probably less rigid than the inner monolayer." A similar conclusion for the LM cell plasma membrane was reached by Schroeder,[12] who used selective quenching of the outer leaflet fluorescence of β-parinaric acid. In contrast to the foregoing studies, however, Tanaka and Ohnishi[13] concluded from ESR studies of human erythrocytes with spin-labeled phospholipids that there is "a more rigid phosphatidylcholine bilayer phase," presumably in the outer leaflet of this cell.

To examine further the question of fluidity asymmetry, to provide new fluorescent probes for assessing the fluidity of each hemileaflet of the human erythrocyte and other cells, and to develop fluorophores that localize and report selectively from the outer hemileaflets of intact cells, we initiated the synthesis of a class of "MIMAR" probes of the general structure MIM-a-R, where MIM is a membrane-impermeant moiety, a is a connecting arm, and R is a lipid-soluble fluorophore. We describe below studies with the first group of these fluorophores, all of which are derivatives containing pyrene as the R group.

Impermeant Pyrene Derivatives

Figure 1, taken from Cogan and Schachter,[14] illustrates the structures of five impermeant derivatives of pyrene. Three compounds have an oligosaccharide as the membrane-impermeant group and two fluorophores utilize the tripeptide glutathione. Variation in the length of the connecting arm of the glutathione derivatives makes it possible to probe different depths of the leaflet. Details of each synthesis and purification have been reported.[14] The oligosaccharide derivatives are prepared by enzymatic oxidation of the C-6 primary alcohol group of the terminal galactose residue in lactose, raffinose, or stachyose. Oxidation to an aldehyde permits subsequent reaction with pyrenebutyryl hydrazide (PBH) to form a Schiff's base, which is reduced with $NaBH_4$. The products are purified by thin-layer chromatography and high-pressure liquid chromatography. The glutathione derivative with a short connecting arm, glutathione-pyrene I (GS-PI), is obtained by reaction of reduced glutathione and *N*-(1-pyrenyl)maleimide at pH 7. The longer connecting-arm derivative, glutathione-pyrene II (GS-PII), is prepared by reaction of glutathione-maleimide I[15] with an excess of ethanedithiol and subsequently with pyrenylmaleimide. The glutathione derivatives are purified by preparative thin-layer chromatography on silica gel plates.

Two experimental methods have proved feasible in applying the foregoing impermeant fluorophores to the estimation of hemileaflet fluidity. In the first procedure intact human erythrocytes and leaky ghost membranes prepared from them are incubated with the probes, and the fluorescence anisotropy of the cell suspensions (outer leaflet loaded) and ghosts (both leaflets loaded) compared. Despite the pres-

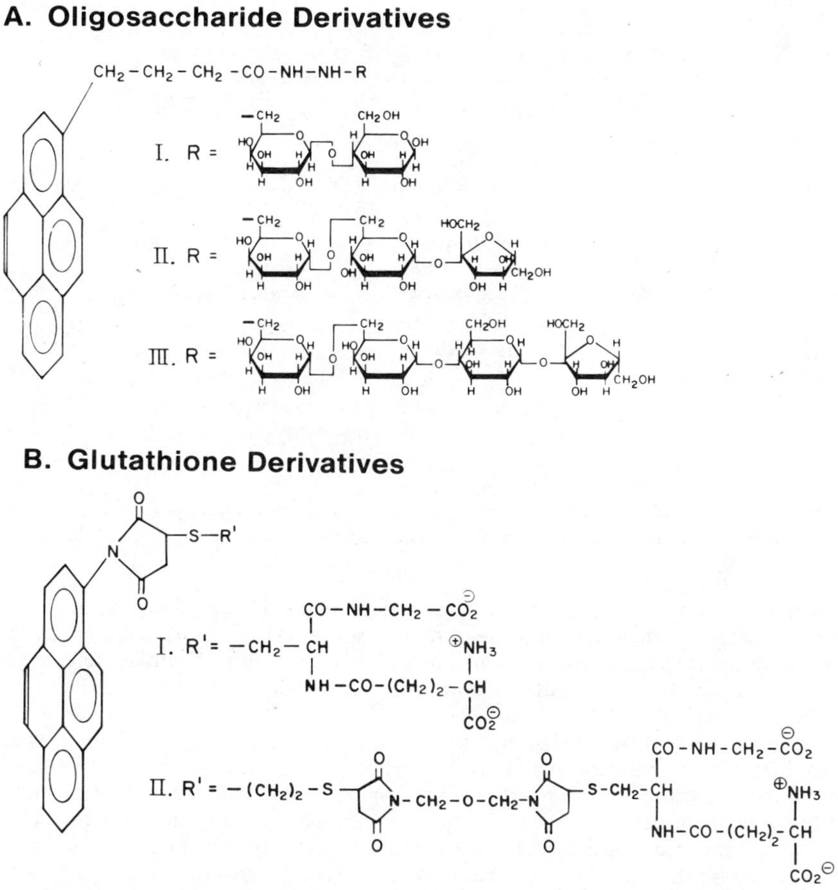

A. Oligosaccharide Derivatives

$CH_2-CH_2-CH_2-CO-NH-NH-R$

I. R =

II. R =

III. R =

B. Glutathione Derivatives

I. R' =

II. R' =

FIGURE 1. Structures of five impermeant derivatives of pyrene. (From Cogan & Schachter.[14] With permission from *Biochemistry*.)

ence of hemoglobin in the cell suspensions, satisfactory fluorescence measurements are possible when the packed cell volume is less than approximately 0.05%.[16] Corrections for scattering are routinely made using controls that lack the probe. Moreover, inasmuch as the excitation and emission spectra of the pyrene fluorophore overlap significantly, concentration-dependent quenching of the fluorescence anisotropy occurs and it is necessary to monitor the molar ratio of probe to lipid in the bilayer in comparative studies. We do this by dissolving the membrane suspensions in 1% (w/vol) sodium dodecyl sulfate and estimating total fluorescence with reference to internal standards added incrementally to each sample. The second procedure to assess hemileaflet fluidity also begins with loading intact erythrocytes (outer leaflets) and ghost membranes (both leaflets). After washing the loaded cells, however, the erythrocytes are lysed osmotically and washed ghost membranes prepared and maintained at 2–5°C until the fluorescence anisotropy is determined. Under these conditions the outer leaflet localization of the cell-loaded preparations is largely maintained for a period of hours, and as shown in TABLE 1,[14] the fluorescence anisotropy values ob-

TABLE 1

FLUORESCENCE ANISOTROPY STUDIES OF ERYTHROCYTE-LOADED AND GHOST-LOADED MEMBRANES

Derivative	Procedure	Preparation Tested	No.	Fluorescence Anisotropy (r)	p
Glutathione-pyrene I	1	intact RBC	3	0.091 ± 0.001	
		ghosts	3	0.147 ± 0.003	<0.001
	2	RBC-loaded membranes	10	0.107 ± 0.005	
		ghost-loaded membranes	10	0.137 ± 0.004	<0.001
Glutathione-pyrene II	1	intact RBC	3	0.044 ± 0.002	
		ghosts	3	0.089 ± 0.005	<0.02
	2	RBC-loaded membranes	24	0.048 ± 0.001	
		ghost-loaded membranes	24	0.074 ± 0.002	<0.001

Values are means ± standard error. Anisotropy values were determined at 25°C.

tained with both procedures above agree well. The particular advantage of the second procedure is that the final membrane suspensions are more comparable and have experienced similar preparations, including osmotic lysis. Hence differences between the outer and inner leaflet fluorescence anisotropy cannot be ascribed to the procedure itself.

The values in TABLE 1 demonstrate a marked difference in fluorescence anisotropy between the hemileaflets. This difference was confirmed with all five impermeant derivatives and related further to an asymmetry of lipid fluidity by estimation of the excited-state fluorescence lifetimes and calculation of the apparent correlation times, assuming no limiting hindered anisotropy.[14] As shown in TABLE 2, the correlation times indicate that the outer leaflet is considerably more fluid than the inner. It is noteworthy that the total probe taken up is considerably less in intact cell-loaded (outer leaflet) preparations as compared to ghost-loaded suspensions (both leaflets). Hence the latter yield fluorescence signals that originate mainly from the inner leaflet. The values in TABLES 1 and 2 also indicate that the membrane microenvironment of glutathione-pyrene I, which has a relatively short connecting arm, is less fluid than that of the remaining probes. This is consistent with the well-documented fluidity gradient of bilayer membranes, in which regions close to the aqueous interfaces are less fluid than those in the hydrophobic core.

The foregoing anisotropy observations characterize hemileaflet fluidity in relation to the rotational diffusion of the pyrene substituent. We were also able to evaluate the short-range lateral diffusion of the fluorophore by the estimation of excimer fluorescence.[14] When erythrocyte-loaded (outer leaflet) and ghost-loaded (mainly inner leaflet) suspensions of ghost membranes were examined, the excimer/monomer ratio was considerably higher in the outer leaflet, indicative of greater fluidity and in agreement with the fluorescence anisotropy studies. This result was obtained with glutathione-pyrene II and the lactose, raffinose, and stachyose derivatives of pyrene-butyryl hydrazide. (Glutathione-pyrene I yielded insufficient excimer for estimation.)

As controls for these experiments with impermeant pyrene derivatives, similar

TABLE 2

APPARENT CORRELATION TIMES OF IMPERMEANT PYRENE DERIVATIVES IN ERYTHROCYTE-LOADED AS COMPARED TO GHOST-LOADED MEMBRANES

Probe	No.	Apparent Correlation Time, τ_c (nsec) RBC-loaded (a)	ghost-loaded (b)	p	Ratio (b)/(a)
Glutathione-pyrene I	6	173 ± 20	311 ± 34	<0.01	1.8
Glutathione-pyrene II	10	32 ± 1	62 ± 3	<0.001	2.0
Lactose-pyrenebutyryl hydrazide	4	23 ± 1	40 ± 1	<0.001	1.7
Raffinose-pyrenebutyryl hydrazide	6	17 ± 2	32 ± 2	<0.001	1.9
Stachyose-pyrene-butyryl hydrazide	4	13 ± 1	27 ± 2	<0.001	2.1

Values are means ± standard errors. Estimations were at 25°C. Correlation times are calculated assuming $r_\infty = 0$.

studies were carried out with membrane-permeant compounds, i.e., pyrene and pyrene decanoic acid. In contrast to the results with the impermeant probes, the fluorescence anisotropy and excimer/monomer intensity ratio observed with erythrocyte-loaded as compared to ghost-loaded membrane suspensions did not differ significantly when tested with fluorophores that can flip from one leaflet to the other.

Since a single fluorophore, such as pyrene, might partition preferentially into selected microdomains within each leaflet, we examined further the asymmetry of fluidity in erythrocytes by using a number of permeant lipid fluorophores: 1,6-diphenyl-1,3,5-hexatriene, 2-(9-anthroyloxy)stearate, 12-(9-anthroyloxy)stearate, and pyrene decanoic acid.[16] Advantage was taken of nonradiative energy transfer from a lipid fluorophore donor to heme acceptors at the endofacial (inner) surface. Since the efficiency of nonradiative energy transfer varies inversely with the sixth power of the distance between donor and acceptor, nonradiative energy transfer is expected to shorten the mean lifetime of donor fluorophores in the membranes of intact erythrocytes. Moreover, in intact cells the excited-state lifetime of the inner leaflet fluorophores should be shortened to a greater extent than that of the outer leaflet probes, and the fluorescence observed should represent disproportionately the outer leaflet fluorophores. Accordingly, we compared the fluorescence anisotropy and mean lifetimes observed for each of the above lipid fluorophores in intact erythrocytes and in lysates.[16] For example, for 2-anthroyloxystearate we could calculate that in lysates (fluorescence from both leaflets) the correlation time at 25°C was 7.5 nsec. On the assumption that this value obtains also for fluorescence from the intact cell suspensions (fluorescence mainly from the outer leaflet), we calculated an expected fluorescence anisotropy value of 0.163 for the intact cell suspensions. The anisotropy observed, however, was 0.154, significantly lower than the predicted value ($p < 0.005$); and similar findings were obtained with all four fluorophores. The results clearly contradict the assumption of identical motional freedom in the two leaflets and support the foregoing conclusion that the outer leaflet is more fluid than the inner.

What is the physiological significance of the asymmetry of hemileaflet fluidity? We have no clear answer as yet, but it has been possible to detect selective perturbations of hemileaflet fluidity and to begin an exploration of the effects of such perturbations on various membrane proteins and their associated functions.

TABLE 3

FLUORESCENCE ANISOTROPY OF STACHYOSE-PYRENEBUTYRYL HYDRAZIDE IN MEMBRANES OF ERYTHROCYTE-LOADED AND GHOST-LOADED ACANTHOCYTES

	Fluorescence Anisotropy (24°C)	
Subject	Erythrocyte-loaded	Ghost-loaded
Patient M. S.	0.025	0.031
Normal A. C.	0.020	0.030
Patient M. S.	0.016	0.027
Normal P. S.	0.012	0.028
Patient A. M. V.	0.014	0.019
Normal P. Y.	0.011	0.019

Patients' values differ significantly ($p < 0.02$) from paired normal control values for erythrocyte-loaded membranes only.

GENETIC AND EXPERIMENTAL PERTURBATIONS CAN ALTER HEMILEAFLET FLUIDITY SELECTIVITY

A congenital human disease (acanthocytosis, or congenital abetalipoproteinemia) and an experimental procedure (modification of the erythrocyte-membrane cholesterol content) can alter the fluidity of the outer hemileaflet selectivity.[17] Acanthocytosis is characterized by abnormally shaped erythrocytes whose membranes contain increased sphingomyelin,[18] and prior studies with diphenylhexatriene had shown that the fluidity of the membrane lipid as a whole is decreased.[18,19] To examine the fluidity of the individual hemileaflets, studies with stachyose-pyrenebutyryl hydrazide were performed as described in the preceding section, in order to compare erythrocyte-loaded and ghost-loaded membrane suspensions. The results in TABLE 3 show that patient's cells yielded higher anisotropy values than did normal cells, but only in the erythrocyte-loaded preparations, indicative of decreased fluidity in the outer leaflet alone. This result is consistent with the known increase in sphingomyelin content of these cells, inasmuch as sphingomyelin is located primarily in the outer leaflet of the erythrocyte membrane[8,9] and has been shown to decrease the fluidity of lipid bilayer membranes.[18] The acanthocyte thus provides an example of a cell altered by disease so that the normal asymmetry of leaflet fluidity is considerably diminished. Given the evidence that changes in the composition of one hemileaflet in re-

TABLE 4

EFFECTS OF CHOLESTEROL ENRICHMENT AND DEPLETION ON HEMILEAFLET FLUIDITY ASSESSED WITH STACHYOSE-PYRENEBUTYRYL HYDRAZIDE

	% Change in Anisotropy Parameter* of Stachyose-Pyrenebutyryl Hydrazide			
Preparation	Erythrocyte-loaded	p	Ghost-loaded	p
Cholesterol enriched	+42.0	(<0.005)	−3.9	(n.s.)
Cholesterol depleted	−14.4	(<0.05)	−21.9	(<0.01)

Values are means ± standard errors for six different blood samples.[17].

* The anisotropy parameter is defined as $[(r_0/r) - 1]^{-1}$ and varies with the correlation time.[14] p values are for differences from control (unmodified) cells.

lation to the other may underly many abnormalities of erythrocyte cell shape,[20] it is reasonable to relate the unusual shapes of acanthocytes to the selective hemileaflet change.

Our studies with stachyose-pyrenebutyryl hydrazide have also demonstrated that modification of the cholesterol content of the erythrocyte membrane can alter hemileaflet fluidity selectively.[17] In particular, cholesterol enrichment leads to decreased fluidity in the outer leaflet alone (TABLE 4). Cholesterol depletion, on the other hand, increases the fluidity mainly of the inner leaflet, with a smaller reduction in the outer leaflet. These results indicate that modulation of membrane cholesterol can provide an experimental tool for exploring the effects of hemileaflet fluidity on membrane protein organization and function.

HEMILEAFLET FLUIDITY AND ERYTHROCYTE MEMBRANE PROTEINS

The working hypothesis we have adopted is that alterations of the lipid fluidity of a specific hemileaflet can lead to changes in organization and/or function of those proteins associated with that leaflet. Inasmuch as the effects of cholesterol modification on erythrocyte-membrane proteins have been studied by a number of laboratories, ours included, it is possible to make an initial test of the hypothesis by an analysis of data already published.

The reactivity of erythrocyte-membrane protein sulfhydryl groups at the exofacial and endofacial surfaces of the membrane, respectively, was examined systematically in relation to cholesterol enrichment and depletion several years ago.[21] The studies were performed by alkylating either intact erythrocytes (outer surface only) or leaky ghost membranes (both surfaces) with a membrane-impermeant sulfhydryl reagent, [^{35}S]glutathione-maleimide.[15,21] These assays provide sulfhydryl group titers for each surface separately, and the values can be related to cholesterol enrichment or depletion. The results in TABLE 5 show that the exofacial sulfhydryl group titer (intact cells alkylated) was significantly affected only by cholesterol enrichment ($p < 0.0025$), with no significant effect of cholesterol depletion. TABLE 6, on the other hand, presents data to show that the endofacial sulfhydryl group titer (ghost membranes alkylated; ~95% of the total surface sulfhydryl groups is endofacial[15]) is affected mainly by cholesterol depletion, with a much smaller effect of cholesterol enrichment. Thus the changes in sulfhydryl group titers support the working hypothesis.

In these studies we were also able to assay the incorporation of [^{35}S]glutathione-maleimide into specific membrane proteins, using sodium dodecyl sulfate-polyacrylamide gel electrophoresis to resolve the proteins labeled by alkylation of ghost mem-

TABLE 5

EFFECTS OF CHOLESTEROL MODIFICATION ON EXOFACIAL SULFHYDRYL TITERS OF INTACT ERYTHROCYTES

Cholesterol/Phospholipid Molar Ratio	Sulfhydryl Labeled (nmol/mg protein)	% Change from Control	p
0.95 (control)	3.0		
0.67 (depleted)	2.9	−3	n.s.
1.19 (enriched)	2.5	−15	<0.0025

Values are means for six different preparations labeled with [^{35}S]glutathione-maleimide.[21] p values are based on paired *t* tests.

TABLE 6

EFFECTS OF CHOLESTEROL MODIFICATION ON MEMBRANE SULFHYDRYL TITERS OF GHOST MEMBRANES

Cholesterol/Phospholipid Molar Ratio	Sulfhydryl Labeled (nmol/mg protein)	% Change from Control	p
0.95 (control)	104 ± 11		
0.67 (depleted)	70 ± 11	−33	<0.01
1.19 (enriched)	115 ± 10	+11	<0.05

Values are calculated from published data[21] and are means ± standard error. The sulfhydryl group titers are essentially endofacial titers, since ∼95% of the membrane surface sulfhydryl groups is endofacial.[15]

branes. The specific radioactivity of each major Coomassie brilliant blue-stained band was quantified by autoradiography and densitometry[21] and the results are listed in TABLE 7. Spectrin, actin, and glyceraldehyde-phosphate dehydrogenase are localized at the endofacial surface, and the values in TABLE 7 show that alkylation of these polypeptides was influenced significantly by cholesterol depletion but not by cholesterol enrichment. (It is noteworthy that cholesterol depletion decreased the sulfhydryl group reactivity of spectrin, actin, and Band 3, but increased the reactivity of glyceraldehyde-phosphate dehydrogenase). The anion transport protein, Coomassie brilliant blue-stained Band 3, spans both leaflets of the membrane, but the sulfhydryl groups labeled in this protein are mainly at the endofacial surface.[15] In this protein cholesterol depletion decreased the sulfhydryl group alkylation by 42.9% ($p < 0.025$), whereas cholesterol enrichment increased the alkylation by only 11.4% ($p < 0.05$). Thus, the working hypothesis is again supported—endofacial proteins and their substituents are affected primarily by cholesterol depletion and not at all, or very little, by cholesterol enrichment.

The counterparts to the endofacial proteins are the exofacial polypeptides, such as the major Rh_o (D) antigen. Several years ago we titered these exofacial sites in cholesterol-modified cells, using [^{125}I]anti-D antibodies,[22] and those results were used to calculate the values in TABLE 8. Here, cholesterol enrichment increased the titer of

TABLE 7

EFFECTS OF CHOLESTEROL MODIFICATION ON SULFHYDRYL REACTIVITY OF INDIVIDUAL MEMBRANE PROTEINS

		% Change in mols Sulfhydryl Groups Labeled per mol Protein	
Protein	Coomassie Brilliant Blue Band	Cholesterol Depletion	Cholesterol Enrichment
Spectrin	1,2	−37.6 ($p < 0.02$)	+6.1 (ns)
Actin	5	−34.7 ($p < 0.005$)	+5.4 (ns)
Anion transport	3	−42.9 ($p < 0.025$)	+11.4 ($p < 0.05$)
Glyceraldehyde-phosphate dehydrogenase	6	+50.0 ($p < 0.025$)	−8.3 (ns)

Mean values calculated from Borochov *et al.*[21] p values are based on paired *t* tests.

TABLE 8

EFFECTS OF CHOLESTEROL MODIFICATION ON THE RH_O (D) ANTIGENIC SITES ON THE EXOFACIAL SURFACE OF THE HUMAN ERYTHROCYTE

	[125I]anti-Rh_O (D) Binding Titer		Apparent K_d of Binding	
Cholesterol/phospholipid Molar Ratio	Sites/Cell	% Difference from Control	Molar	% Difference from Control
0.9 ± 0.1 (control)	5,000–6,000		2.5×10^{-10}	
0.6 ± 0.1 (depleted)	3,000–3,500	−41	1.7×10^{-10}	−32
1.5 ± 0.1 (enriched)	12,000–14,000	+136	5.0×10^{-10}	+100

Values are calculated from the results of Basu *et al.*[22]

exofacial sites by 136% and the apparent K_d of binding by 100%. By comparison, cholesterol depletion decreased the titer by only 41% and the apparent K_d by 32%. Thus a protein associated with the outer leaflet is influenced mainly by cholesterol enrichment, again in conformity with the working hypothesis.

Additional observations on certain functional proteins, which are not yet as well characterized as the polypeptides mentioned above, also suggest that selective alterations of hemileaflet fluidity can influence specific functions. For example, the facilitated diffusion of D-glucose across human erythrocyte membranes is mediated by a trans-membrane polypeptide of molecular weight ~55,000 and possibly other proteins.[23–25] Recently we have observed that the half-saturation concentration of D-glucose binding to ghost membranes is significantly increased by cholesterol enrichment but not by cholesterol depletion. In view of the working hypothesis developed above, it is reasonable to propose that the glucose-transport mechanism is sensitive to a change in fluidity of the outer leaflet specifically, possibly because a rate-limiting transport event occurs in that locale.

SUMMARY AND CONCLUSIONS

The impermeant fluorescent probes (MIMAR reagents) described here permit the assessment of the lipid fluidity of individual membrane hemileaflets. They should also prove useful for examining the outer hemileaflets of the plasma membranes of intact cells. The observations, thus far, that normal human erythrocyte membranes have a characteristic asymmetry of fluidity, with the outer leaflet more fluid, correspond to prior findings with *Mycoplasma*,[10] Newcastle Disease viral envelopes,[11] and mouse LM cells.[11,12] Hence, it is possible that the pattern is quite general in biological membranes. The particular lipid and protein components of the human-erythrocyte membrane that underly the fluidity asymmetry are unknown. The increased content of phosphatidylcholine in the outer leaflet and of the anionic phospholipids in the inner leaflet would be consonant with the fluidity difference. On the other hand, sphingomyelin, which tends to decrease fluidity, is localized mainly in the outer leaflet. Unknown at present is whether the cholesterol content of the two leaflets differs. From the results reported above, it is tempting to speculate that exogenously added cholesterol tends to localize in the outer leaflet, normally the more fluid leaflet, whereas endogenous cholesterol is more readily removed from the inner leaflet. This suggests, but clearly does not establish, that in the normal erythrocyte the cholesterol content of the inner leaflet exceeds that of the outer. Lastly, integral membrane proteins are expected to decrease lipid fluidity, and the usual pattern seen on freeze-fracture of large numbers of intra-membranous particles on the cytoplasmic face may signify a greater influence of protein in the inner leaflet.

The hypothesis that perturbations of the fluidity of a given hemileaflet influence the membrane proteins (and their associated functions) in that leaflet is well-supported by the evidence described above. On the other hand, we understand less well the mechanisms by which lipid fluidity influences the proteins. For example, the decrease in sulfhydryl group reactivity of spectrin, actin, and Band 3 owing to cholesterol depletion (TABLE 7) may be due to a physical displacement of these proteins, as suggested by Borochov and Shinitzky.[26] Why then does the reactivity of glyceraldehyde-phosphate dehydrogenase sulfhydryl groups increase under these conditions? There remains much to learn about membrane molecular mechanics and lipid-protein interactions. In such studies the impermeant MIMAR probes described here should prove useful.

REFERENCES

1. CHEN, L. A., R. E. DALE, S. ROTH & L. BRAND. 1977. J. Biol. Chem. **252**: 2163–2169.
2. HILDENBRAND, K. & C. NICOLAU. 1979. Biochim. Biophys. Acta **553**: 365–377.
3. LACKOWICZ, J. R., F. G. PRENDERGAST & D. HOGEN. 1979. Biochemistry **18**: 508–519.
4. LACKOWICZ, J. R., F. G. PRENDERGAST & D. HOGEN. 1979. Biochemistry **18**: 520–527.
5. HEYN, M. P. 1979. FEBS Lett. **108**: 359–364.
6. JÄHNIG, F. 1979. Proc. Natl. Acad. Sci. USA **76**: 6361–6365.
7. VAN BLITTERSWIJK, W. J., R. P. VAN HOEVEN & B. W. VAN DER MEER. 1981. Biochim. Biophys. Acta **644**: 323–332.
8. ROTHMAN, J. E. & J. LENARD. 1977. Science **195**: 743–753.
9. OP DEN KAMP, J. A. F. 1979. Annu. Rev. Biochem. **48**: 47–71.
10. ROTTEM, S. 1975. Biochem. Biophys. Res. Commun. **64**: 7–12.
11. WISNIESKI, B. J. & K. K. IWATA. 1977. Biochemistry **16**: 1321–1326.
12. SCHROEDER, F. 1978. Nature (London) **276**: 528–530.
13. TANAKA, K.-I. & S.-I. OHNISHI. 1976. Biochim. Biophys. Acta **426**: 218–231.
14. COGAN, U. & D. SCHACHTER. 1981. Biochemistry **20**: 6396–6403.
15. ABBOTT, R. E. & D. SCHACHTER. 1976. J. Biol. Chem. **251**: 7176–7183.
16. SCHACHTER, D., U. COGAN & R. E. ABBOTT. 1982. Biochemistry **21**: 2146–2150.
17. FLAMM, M. & D. SCHACHTER. 1982. Nature (London) **298**: 290–292.
18. COOPER, R. A., J. R. DUROCHER & M. H. LESLIE. 1977. J. Clin. Invest. **60**: 115–121.
19. BARENHOLZ, Y., E. YECHIEL, R. COHEN & R. DECKELBAUM. 1981. Cell Biophys. **3**: 115–126.
20. SHEETZ, M. P. & S. J. SINGER. 1974. Proc. Natl. Acad. Sci. USA **71**: 4457–4461.
21. BOROCHOV, H., R. E. ABBOTT, D. SCHACHTER & M. SHINITZKY. 1979. Biochemistry **18**: 251–255.
22. BASU, M. K., M. FLAMM, D. SCHACHTER, J. F. BERTLES & A. MANIATIS. 1980. Biochem. Biophys. Res. Commun. **95**: 887–893.
23. BATT, E. R., R. E. ABBOTT & D. SCHACHTER. 1976. J. Biol. Chem. **251**: 7184–7190.
24. KASAHARA, M. & P. C. HINKLE. 1977. J. Biol. Chem. **252**: 7384–7390.
25. CARTER-SU, C., J. E. PESSIN, R. MORA, W. GITOMER & M. P. CZECH. 1982. J. Biol. Chem. **257**: 5419–5425.
26. BOROCHOV, H. & M. SHINITZKY. 1976. Proc. Natl. Acad. Sci. USA **73**: 4526–4530.

MODIFICATION OF CELL MEMBRANE COMPOSITION BY DIETARY LIPIDS AND ITS IMPLICATIONS FOR ATHEROSCLEROSIS

Fred A. Kummerow

Burnsides Research Laboratory
University of Illinois at Urbana-Champaign
Urbana, Illinois 61801

Modification of cell membrane composition has been considered a causative factor in the development of many diseases, including atherosclerosis. Membranes consist primarily of protein and lipid (phospholipids and cholesterol), and secondarily of carbohydrate. Membranes are essential cellular components that envelop the cell and divide it into various functional compartments. They maintain a permeability barrier between these compartments, transmit information, and allow the selective movement of nutrients across them. The phospholipids, through the hydrophobic character of the fatty acids attached to them, constitute the predominant structural element of membranes and impart to them an impermeability to hydrophilic molecules. The carbohydrate groups are attached to lipids or proteins and are localized mainly at the cell surface. They play a vital structural role in determining receptor and antigenic sites. Proteins may be either peripherally associated with membranes or may be an integral component whereby they usually span the membrane. They function as recognition or catalytic sites. A change in any of these essential structural components may precipitate a disease process. Increasing significance is now being attached to the role of membrane lipids, both directly in influencing membrane properties and indirectly through their effect on membrane proteins.

In our laboratory we are primarily concerned with the role of dietary lipids in nutrition and their possible role in initiating disease processes, particularly atherosclerosis. The molecular mechanisms underlying the changes inflicted by these dietary lipids all appear to involve membrane-related phenomena and in many cases may stem from their incorporation into membranes. In this presentation I will relate possible membrane changes due to dietary fatty acids, oxidized sterols, or vitamin D.

MEMBRANES AND DISEASES

In view of the important functions of membranes, it is not surprising that altered functions are being linked to the initiation and development of disease processes, and serving as diagnostic evidence of these processes. In TABLE 1, I have ranked diseases in order of causes of death and have related these diseases to membrane functions that have been proposed to be involved. The numerical references (1–40) listed in each column are examples of articles that have examined the association of these diseases with membrane functions. This list is very superficial and it is very likely that a more exhaustive search of the literature would enable each box to be filled.

I have divided the areas of research on cell membranes in the disease process into two functional spheres of influence: the functions related to the protein components of the membrane and the functions related to the lipid phase of the membrane. The glycoproteins attached to the cell membrane could be considered to function as

0077-8923/83/0414-0029$01.75/0

TABLE 1

FUNCTIONAL CHANGES IN ADHERING PROTEINS, RECEPTOR SITES, AND LIPID PHASES IN THE DISEASE PROCESS

	Protein Functions						Lipid Phase		
Disease Process	Antigenic recognition	Intracellular communication	Permeability (Ion Transfer)	Receptor Interaction	Enzyme Activities	Endocytosis	Sterols	Fatty Acids	Cell Fusion
Heart			25, 26	39	4		13, 27, 31	28, 30, 32	
Cancer	1, 2, 24	24	2, 32	1, 2, 3, 15	1, 14			32	2
Cerebrovascular	1			1					
Influenza (viruses)	17	16		17, 20					18, 21
Diabetes	38	38	1, 38	38	38		38	38	
Liver			29	37	37		11, 13	12, 29, 37	
Muscular dystrophy			1, 7	1, 10	1, 7, 19, 40		7	7, 22	
Sickle cell			1, 2, 8			1	9		
Red cell membrane	8, 15		1, 8, 23, 25	15		1, 13			1
Parasites	5		2, 6	5					
Affective disorders	33	1						35	
Gastrointestinal			1, 34		1			34	
Kidney	36		1, 36	36					

antigenic sites and, in intracellular communications, as receptors for hormones and growth factors. Receptor sites may be linked to membrane-bound enzymes or transport proteins and may be endocytosed. The lipid phase is the main structural entity of the membrane and provides the permeability barrier to compartmentalize cellular components. It may play a key role in membrane fusion and membrane permeability as well as influencing the activity of membrane-bound proteins. These two spheres of influence seem to overlap or commingle so that their specific involvement in the disease process cannot be differentiated without further study.

An example of the involvement of the cell surface in membranes is provided in studies on cancer as well as heart disease. Membranes contain in their surface coats communication systems that keep the individual cells in a tissue or organ functioning in a cooperative manner. The communication system recognizes neighboring cells and constrains the capacity of cells to proliferate. This system may involve the glycolipids and glycoproteins that are attached to the cell surface. Wallach[2] has pointed out that malignant cells do not appear to complete the carbohydrate components of their glycolipids and may thus develop an incomplete communication system allowing their uncontrolled proliferation.[1-3] Some investigators[31-33] believe that the development of an atherosclerotic plaque results from the migration of medial smooth muscle cells into the intima, most likely in response to a chemoattractant, such as platelet-derived growth factor. This would involve a surface interaction between the growth factor and receptor. Lipid accumulation in these migrated smooth muscle cells may then result from receptor-mediated endocytosis of low density lipoproteins and the inability of the cells to metabolize the cholesterol in low density lipoproteins. However, factors other than an excessive amount of cholesterol may be crucial to the development of atherosclerosis, as I will discuss later. The cell surface also seems to be involved in abnormalities of the red blood cell, including sickle cell anemia[8,9] and muscular dystrophy.[7,10]

The linkage of membranes to disease has encountered three difficulties. One, significant change in membrane properties may be difficult to detect because of the lack of morphological change accompanying the functional change, and the difficulty in technically measuring some membrane properties. Two, isolating the membrane in a pure form that reflects its *in vivo* functional properties is sometimes difficult. Three, it is often not clear whether a change in membrane properties is a result of the disease, or whether it truly represents an active role in the initiation and progression of the disease. The application of magnetic resonance, fluorescence polarization, laser technology, electron microscopy, and other sophisticated techniques to this question, as is evident from other contributions to this volume, may clarify these issues.

Fatty Acids and Sterols in Membrane Function

The fatty acids that are acylated to glycerol in phospholipids are key structural elements in membranes. Much smaller amounts of fatty acids are esterified to other lipid classes or to proteins, or exist free in the membrane. To form a membrane the fatty acid hydrocarbon chain must be of a length within certain limits, generally containing between 12 and 24 carbon atoms. Biological membranes in general contain acids with 16, 18, or 20 carbon atoms. Sphingomyelin contains a large percentage of fatty acids, 24 carbon atoms in chain length, and retinal membranes contain a large percentage of C22:6, indicating that in some instances longer chain acids are important to membrane structure. This range of fatty acids creates membranes with a restricted bilayer thickness. Recent experiments with a purified membrane protein, the sarcoplasmic reticulum Ca^{2+}-ATPase, reconstituted in liposomes of different composition, indicated that optimal activity was obtained when the membrane phospholipids contained fatty acids with 20 carbon atoms.[41] This reflects the functional importance of acyl chain length and the resultant bilayer thickness, for optimal enzymatic activity.

Fatty acid unsaturation is another key factor related to membrane function. Its principal effect appears to be to create fluid environments in the membrane so that, proteins and lipids can diffuse in the plane of the membrane, and protein can undergo a conformational change during interaction with a ligand, substrate, or effector. Experiments with model membranes, fatty acid auxotrophs of *E. coli*, and cultured cells *in vitro*, have shown that fatty acids can influence phase-transition temperatures of membrane lipids and the activities of membrane-bound enzymes and transport proteins.[42,43]

A factor, which has largely been ignored to date is the presence of isomeric fatty acids in the phospholipids isolated from human heart, aorta, and liver tissue, presumably as structural components of their biological membranes.[44,45] Isomeric fatty acids are present in hydrogenated fats that have been introduced into the food chain as shortenings and margarine as a replacement for lard, tallow, and butter, and are present in a wide variety of food products. They are produced by catalytic hydrogenation of vegetable or fish oils. In the process there is a migration of some double bonds along the acyl chain in unsaturated fatty acids, producing positional isomers of *cis* orientations. Geometric isomers of *trans* orientations may also form during the regeneration of double bonds during hydrogenation.

The double bond in positional isomers of monoenoic acids may range from the Δ6 to the Δ13 position with the Δ9, Δ10, and Δ11 being the main isomers.[46] Small percentages of isomers of dienoic acids may also be produced. Isomerization changes the spatial configuration so that the *trans* form of oleic acid (elaidic) acid displaces a greater area than stearic acid but less than oleic acid (FIGURE 1). This change in spatial configurations and the tighter packing of *trans* monoene acyl chains compared with *cis* monoene acyl chains is also reflected in their melting points. Thus, elaidic acid has a melting point as well as a molecular volume intermediate between

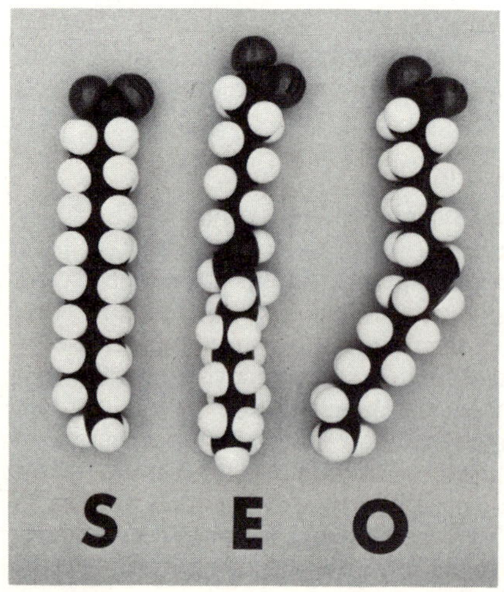

FIGURE 1. Spatial configuration of stearic (S) (m.p. 69°C), elaidic (E) (*trans*) (52°C), and oleic (O) (*cis*) (14°C) acids.

those of stearic and oleic acids. The influence of these isomeric fatty acids on membrane function remains to be clarified.

Enzymes, which have to discriminate between fatty acids based on precise positions and geometric forms of double bonds are, understandably, somewhat confused by these isomers so that acyltransferases treat monoenoic *trans* unsaturated isomers in a manner similar to saturated fatty acids and insert them at the α-position of phospholipids. It has been shown by Lands *et al.*[47,48] that the monoenoic *trans* form usually replaces a saturated fatty acid at the α-position and Kinsella *et al.*[49] have shown that dienoic *trans,trans* form can replace dienoic *cis,cis* form at the β-position. Positional isomers are not good substrates for desaturases, although there are reports that linoelaidic acid can be elongated and desaturated to form a 20:4 isomeric fatty acid.[50]

Cholesterol is the predominant sterol in animal cell membranes and its main functional role identified to date appears to be in the modification of the motional freedom of fatty acyl chains.[51,52,90] When added to membranes consisting of lipids with a gel-like molecular arrangement, it increases their motional freedom. Conversely, it reduces the mobility of acyl chains in a fluid state.[53] Structural features of cholesterol required to produce this effect include a 3β-OH group, a planar steroid nucleus, and an isoprenoid side-chain of precise length and saturation.

The Influence of Dietary Fatty Acids and Sterols on Membrane Function and Its Relationship to Atherosclerosis

Fatty Acids

As indicated in Table 1, atherosclerosis and arteriosclerosis, the causes of coronary heart disease and strokes, respectively, are the number one cause of death in practically every technologically advanced country in the world. The relationship of dietary fatty acids to coronary heart disease has been considered from two major perspectives: the extent of fat in the diet and the saturation of the dietary fatty acids. Other considerations involving dietary fatty acids include the presence of isomers,[54] the oxidation of polyunsaturated acids,[55] and the presence of long chain fatty acids, such as erucic acid.[56]

The major concern of clinical investigators has been the effect of saturation of dietary fats on serum cholesterol levels. There may be a correlation between serum cholesterol and coronary heart disease, but, to date, there has been no causative relationship linking dietary cholesterol levels with heart disease. As I will discuss in greater detail later, the lack of relationship has been shown in experiments where calcium blockers and antagonists, such as lanthanum[57,58] and nifedipine,[59] prevented the appearance of extensive lesions in animals fed an atherogenic, cholesterol-rich diet. Rabbits in these experiments had serum cholesterol levels in excess of 2,000 mg/100 ml serum.

The direct effect of the saturation of dietary fatty acids on the properties of arterial cell membranes has largely been ignored. It has been proposed that alterations in membrane fluidity may underlie the development of artherosclerosis.[39,60] These proposals have centered on variations in membrane cholesterol as being the critical factor influencing membrane fluidity. Equally, however, a change in membrane fatty acid saturation could affect membrane fluidity and deserves consideration.

When viewing the role of dietary fatty acids in influencing membrane properties, the presence of several different classes of fatty acids must be considered. These include: (a) the presence or absence of essential fatty acids; (b) the relative amounts of saturated and unsaturated fatty acids; (c) the presence of isomers; and (d) the relative amounts of ω6 and ω3 fatty acids. The relationship of (a) and (d) to membrane function does not appear to be direct but involves the synthesis of eicosanoids such as the thromboxanes, prostaglandins, and leukotrienes. These in turn may influence membrane properties but their mode of action is not yet certain. Although it has been assumed that polyunsaturated fatty acids may be essential membrane components,[61] there is no evidence to support this assumption. In fact several cell lines can now be grown in culture in serum-free media lacking any exogenous fatty acids.[62] The membranes of these cells apparently contain only trace amounts of polyunsaturated fatty acids of the ω7 series. Thus, this raises doubts as to the requirements for polyunsaturated fatty acids for cellular membrane function *in vivo*, although there is no question that essential fatty acids are required for the synthesis of eicosanoids for the maintenance of tissue response in physiological functions.

TABLE 2

PUBLISHED EXPERIMENTS PERTAINING TO THE SATURATION AND ISOMERIZATION OF DIETARY FATS

Experimental Animal	Comparative Diets (conc.)	Membrane Function	Comments	Reference
Sheep	Sunflower oil (?) vs. pasture (?)	Mitochondrial succinate oxidation	Type of diet and levels of fat different	65
Rat	Sunflower oil-soaked pellets (?%) vs. pellets (?%)	Mitochondrial succinate oxidation	Level of fat different	65
Rat	Corn oil (15%) vs. hydrogenated soy bean oil (15%)	Mitochondrial respiration	Hydrogenated oil deficient in essential fatty acids	66
Rat	Olive oil (8%) vs. elaidinized olive oil (8%)	Mitochondrial & erythrocyte permeability	Elaidinized olive oil deficient in essential fatty acids & olive oil marginally deficient	67
Rat	Sunflower oil (14%) vs. coconut oil (14%)	Microsomal enzymes	Coconut oil deficient in essential fatty acids	68
Mouse	Sunflower oil (14%) vs. coconut oil (14%)	Plasma membrane amino acid transport	Coconut oil deficient in essential fatty acids	69
Rat	Soybean oil (20%) vs. rapeseed oil (20%)	Mitochondrial oligomycin-sensitive ATPase	Rapeseed oil contains a high level of erucic acid, which causes pathological effects	70
Rat	Hydrogenated lard (5%) vs. lard (5%) vs. corn oil (5%) vs. linseed oil (5%)	Erythrocyte Mg^{2+}-ATPase	First two diets are EFA-deficient containing less than 2% as caloric intake. Deficient in Zn^{2+} and Mn^{2+}	71

Experiments with model membranes, prokaryotes, and cultured cells have confirmed that the saturation of membrane fatty acids can influence membrane properties. However, *in vivo* experiments attempting to relate the degree of saturation of dietary fatty acids to a resultant change in membrane properties in mammals have not yielded clear-cut results. This difficulty seems to relate to the complex pathways of metabolism of dietary fats and the regulation of many of these pathways to avoid large changes in membrane lipid composition. Moreover, many dietary experiments have not considered the contribution of the eicosanoids arising from essential fatty acids to membrane properties. The well-recognized changes in membrane function as a result of essential fatty acid deficiency attest to this.[63,64] An example of membrane changes determined in our laboratory is described in Holmes *et al.*[91] An examination of many of the published experiments pertaining to the effect of the saturation and isomerization of dietary fatty acids shows that they suffer from experimental deficiencies that do not allow an unequivocal interpretation of experimental results (TABLE 2). In such experiments, the type or level of dietary fatty acids may have varied between experimental treatments, the sufficiency of essential fatty acids may have varied, and the diets may have been deficient in trace minerals, which can influence lipid metabolism. The effect of these parameters, level, type, essential fatty acids (EFA) deficiency, and trace mineral deficiency on membrane composition, membrane properties, or lipid metabolism, have been documented.[65-71] Despite the reservations concerning the interpretation of these experiments, the observable trend in these experiments is that dietary fatty acid saturation influences membrane properties. A range of membrane functions are affected including membrane permeability and the activity of membrane-bound enzymes. However an unequivocal interpretation awaits better nutritionally designed experiments.

Changes in membrane fatty acid composition have been detected in the membranes of experimental animal models fed various fats. While attempts have been made to correlate directly an increase in dietary fatty acid saturation with an increase in the saturation of membrane fatty acids, this effect has not been observed in many experiments. It is apparent that the membrane regulates to some extent its fatty acid composition in an attempt to control the fluidity of membrane lipids. Examples of membrane-lipid changes in studies carried out in our laboratory are listed in TABLES 3 and 4. A higher percentage of linoleic acid was noted in the erythrocyte lipid of swine fed corn oil as compared to those fed beef tallow. A similar trend was noted in the lipids extracted from the heart sarcolemma membrane of swine fed corn oil as compared to those fed hydrogenated fat. The cholesterol content was similar in both cases, but the mixed fatty acid composition was not.[72,73]

The effect of isomeric fatty acids is not clear-cut for the same reasons. In our

TABLE 3

CHOLESTEROL AND FATTY ACID CONTENT OF ERYTHROCYTE LIPIDS OBTAINED FROM SWINE FED 30% OF CALORIES FROM BEEF TALLOW OR CORN OIL

	Beef Tallow	Corn Oil
Cholesterol mg/100 ml	135	136
Palmitic C16:0(%)	35	31
Stearic C18:0	13	15
Oleic C18:1	31	20
Linoleic C18:2	16	29

(After Kummerow *et al.*[72])

TABLE 4

CHOLESTEROL AND FATTY ACIDS IN PHOSPHATIDYLCHOLINE OF CARDIAC SARCOLEMMA MEMBRANE ISOLATED FROM SWINE FED 18% OF CALORIES FROM HYDROGENATED FAT OR CORN OIL

	Hydrogenated Fat	Corn Oil
Sarcolemma Cholesterol (mg/mg lipid)*	0.204	0.182
Palmitic C16:0(%)	29.3	29.3
Stearic C18:0	6.5	8.1
Oleic C18:1	30.2†	16.3
Linoleic C18:2	25.9	33.0
Arachidonic C20:4	5.5	5.6

* Not significant.
† Contains 2.9% *trans*; OV-275 column.
(After Babka.[73])

laboratory we are attempting to analyze how dietary fatty acid saturation and isomerization influence membrane properties in carefully controlled experiments.[91]

Sterols and Oxidized Derivatives

As mentioned earlier, a great deal of attention has been focused on both the relationship of dietary cholesterol and serum cholesterol to the development of heart disease. However, as I have stated previously, such relationships are tenuous. In terms of altered membrane functions, cholesterol levels in the diet and serum could affect the amount of cholesterol in the membrane. The changes observed in membrane cholesterol both *in vivo* and *in vitro* over a range of dietary and serum levels suggest that significant changes do not occur under normal physiological conditions. Thus, while proposals that changes in membrane fluidity occur in response to elevated serum cholesterol levels appear attractive, they are not supported by experimental evidence.

A proposal we have been examining in our laboratory is that oxidized sterols may be a more significant factor than cholesterol in influencing the initiation and development of atherosclerosis through their effect on membranes. 25-Hydroxycholesterol and 7-ketocholesterol[74-76] are auto-oxidation products of cholesterol and may arise during processing or cooking of cholesterol-containing foods. The potential for these derivatives to exist in the food chain and to produce pathological effects may be ascertained by performing some simple calculations. If the average American consumes 500 mg of cholesterol per day and if 0.1% of that were oxidized (a very small proportion), this would result in the consumption of 500 μg of oxidized derivatives of cholesterol. If 20% of that were 25-hydroxycholesterol, it would be equivalent to the ingestion of 100 μg 25-hydroxycholesterol per day. We believe that the toxicity of vitamin D is related to the formation of 25-hydroxyvitamin D, the main circulating form of vitamin D, which shares several common structural features with 25-hydroxycholesterol. Toxic responses to vitamin D[77] have been reported with doses as low as 50 μg per day, which stresses the relevance of cholesterol oxidation. However, uncertainties exist as to the actual extent of formation of oxidized products, their rate of absorption, and their rate of metabolism.

We have found[78] that 7-ketocholesterol is quite potent in increasing the frequency of degenerated smooth muscle cells in the aortas of experimental chicks (FIG-

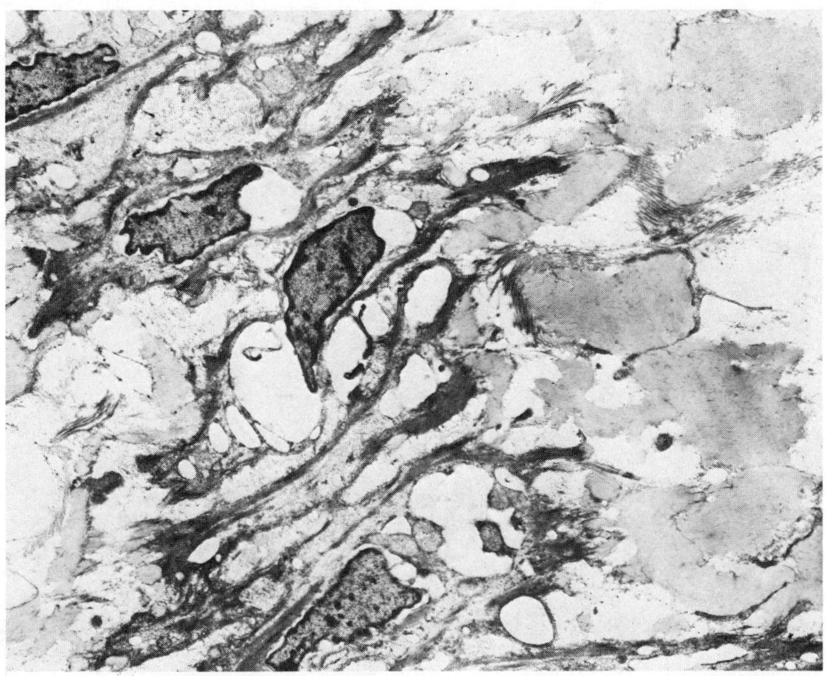

FIGURE 2. Thoracic aorta from a chick fed 0.5 ml 2% 7-ketocholesterol in corn oil for 8 weeks (×720).

URE 2). Other workers have confirmed that 25-hydroxycholesterol is similarly angiotoxic.[79] In our studies on chicks we were not able to determine whether significant Ca^{2+} deposition occurred in the coronary arteries of those fed 7-ketocholesterol, as the techniques employed would have permitted leaching of Ca^{2+} from deposition sites. The question of maintaining Ca^{2+} deposition in thin sections in a way that reflects *in situ* deposition is one being addressed in several laboratories, but with no clear resolution as yet. We are pursuing the hypothesis that one of the principal toxic effects of these sterols is caused by their insertion into cellular membranes. This would influence the permeability properties of membranes, particularly in the permeability of plasma membranes to Ca^{2+}, as a large electrochemical gradient for Ca^{2+} exists across these membranes.

Similar principles may apply during hypervitaminosis D where 25-hydroxyvitamin D is the main circulating metabolite.[80] It too may influence the permeability properties of membranes following its insertion in the membrane. Arterial calcification and smooth muscle cell necrosis are typical features of hypervitaminosis D (FIGURE 3).

Hydroxysterols may also influence membrane properties indirectly in two ways. First, they are well-known inhibitors of the synthesis of sterols and other lipid products that are derived from hydroxymethylglutaryl-coenzyme A reductase activity.[81] The flow of sterols and other lipids through cellular membranes may be inhibited, although sufficient cholesterol to meet membrane requirements should be provided by serum cholesterol. Second, 1,25-dihydroxyvitamin D has been shown to increase the fluidity of intestinal brush border membranes by increasing the relative incorpora-

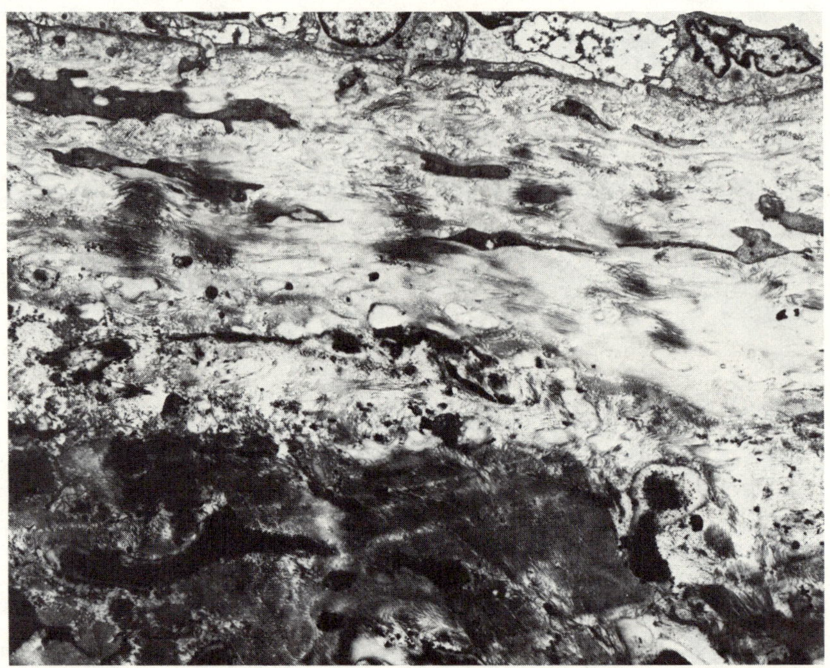

FIGURE 3. Thoracic aorta from 6-month-old swine fed an excess amount of vitamin D_3 for three months (×2,790).

tion of polyunsaturated fatty acids into phospholipids and by increasing the amount of phosphatidylcholine in the membrane.[82-84] This increase in fluidity has been proposed to open cryptic Ca^{2+} channels in the membrane, stimulating Ca^{2+} flow into the cell (FIGURE 4). In view of the increasing number of tissues being identified as targets of 1,25-dihydroxyvitamin D action, [85-88] it is possible that arterial cells are also influenced. If so, during hypervitaminosis D, the high levels of 25-hydroxyvitamin D may simulate 1,25-dihydroxyvitamin D action. One of the responses in arterial cells may be to influence membrane fluidity and Ca^{2+} flow into the cell. This is presently under investigation in our laboratory.

The importance of calcium in the development of atherosclerosis has been shown in studies with calcium antagonists. Kramsch *et al.*[58] found that rabbits fed the calcium antagonist, lanthanum trichloride (40 mg/kg body weight), and 2% cholesterol, did not develop lesions to the same extent as those fed 2% cholesterol alone despite their serum cholesterol levels being similar (2,431 and 2,573 mg %, respectively). Similar results were noted in rabbits fed 2% cholesterol and the calcium "blocker," nifedipine. These studies indicate that the factors that control the flow of calcium through the cell membrane, rather than the presence of an excessive amount of LDL cholesterol[89] in the serum, are the crucial factors in the development of atherosclerosis.

SUMMARY

Dietary lipids can modify the properties of cell membranes, including membrane fluidity and membrane permeability. The saturation and isomerization of di-

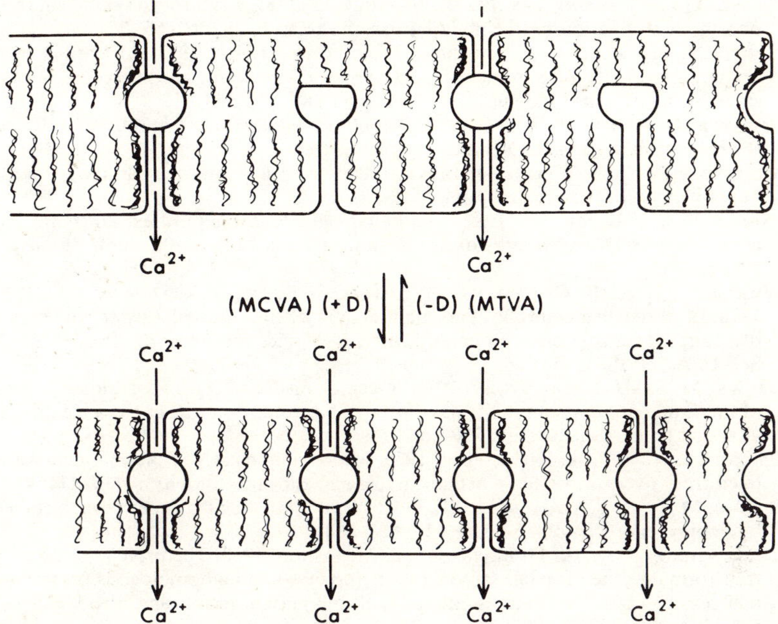

FIGURE 4. A possible model of action suggested by Rasmussen *et al.*[83] by which the methyl esters of *cis*- and *trans*-vaccenic acid (C18:1, 11Δ) (MCVA and MTVA) added *in vitro* produce changes in calcium transport across the luminal membrane of the enterocyte similar to those seen after administration of 1,25-$(OH)_2$-D_3 *in vivo*. In the proposed model, either manipulation (MCVA *in vitro* or 1,25-$(OH)_2$-D_3 *in vivo*) leads to an increase in membrane fluidity, which in turn converts cryptic calcium channels in the membrane into active ones. Conversely, addition of MTVA *in vitro* reverses the effect of 1,25-$(OH)_2$-D_3 by decreasing membrane fluidity. (From Rasmussen *et al.*[83] By permission of *Federation Proceedings*.)

etary fatty acids may affect the pattern of fatty acids acylated to glycerol in phospholipids. Oxidized sterols may affect membrane properties directly by their insertion in the membrane or indirectly through their effects on lipid metabolism. The flow of calcium and other nutrients into the cells appears to be a major property affected by those changes in lipid composition of membranes and may be important in the onset of atherosclerosis. The factors that alter the character of the lipids in cell membranes should receive increased study in both *in vitro* and *in vivo* systems to clarify their role in diseases processes.

REFERENCES

1. BOLIS, L., J. F. HOFFMAN & A. LEAF, EDS. 1976. Membranes and Disease. Raven Press. New York.
2. WALLACH, D. F. H. 1973. The role of the plasma membrane in disease processes. Biol. Membr. **2**: 253–293.
3. VOLKOV, E. I. & D. S. CHERNAVSKI. 1981. Biological consequences of the physical organization of the plasma membranes of normal and tumor cells. Biol. Bull. Acad. Sci. USSR **8**(1): 21–32.
4. GODIN, D. V., J. M. TUCHEK & M. MOORE. 1980. Membrane alternation in acute myocardial ischemia. Can. J. Biochem. **58**(10): 777–786.

5. Pasvol, G., J. S. Wainscoat & D. J. Weatherall. 1982. Erythrocytes deficient in glycophorin resist invasion by the malarial parasite. Nature **297**: 64–66.
6. Kutner, S., D. Baruch, H. Ginsburg & Z. I. Cabantchik. 1982. Alterations in membrane permeability of malaria-infected human erythrocytes are related to the growth stage of the parasite. Biochim. Biophys. Acta **687**: 113–117.
7. Plishker, G. A. & S. H. Appel. 1980. Red blood cell alterations in muscular dystrophy: the role of lipids. Muscle Nerve **3**: 70–81.
8. Mohandas, N., R. I. Week & M. Bessis. 1980. Red cell membrane pathology. *In* Pathobiology of Cell Membranes. **2**: 41–89.
9. Jain, S. K. & S. B. Shohet. 1982. Red blood cell (^{14}C)cholesterol exchange and plasma cholesterol esterifying activity of normal and sickle cell blood. Biochim. Biophys. Acta **688**: 11–15.
10. Wilkerson, L. S., R. C. Perkins, R. Roelofs, L. Swift, L. R. Dalton & J. H. Park. 1978. Erythrocyte membrane abnormalities in Duchenne muscular dystrophy monitored by saturated transfer electron paramagnetic resonance spectroscopy. Proc. Natl. Acad. Sci. USA **75**(2): 838–841.
11. Brown, M. S. & J. L. Goldstein. 1976. Receptor-mediated control of cholesterol metabolism: Study of human mutants has disclosed how cells regulate a substance that is both vital and lethal. Science **191**: 150–154.
12. Perkins, R. G. & F. A. Kummerow. 1976. Major lipid classes in plasma membrane isolated from liver of rats fed a hepatocarcinogen. Biochim. Biophys. Acta **424**: 469–480.
13. Brown, M. S., P. T. Kovanen & J. L. Goldstein. 1981. Regulation of plasma cholesterol by lipoprotein receptors. Science **121**: 628–635.
14. Courtneidge, S. A., A. D. Levinson & J. M. Bishop. 1980. The protein encoded by the transforming gene of avian sarcoma virus (pp 60^{src}) and a homologous protein in normal cells (pp $60^{proto\text{-}src}$) are associated with the plasma membrane. Proc. Natl. Acad. Sci. USA **77**(7): 3783–3787.
15. McNutt, N. S. & S. Hoffstein. 1982. Membranes and cytoskeleton: Role in pathologic processes. Fed. Proc. **40**(2): 206–213.
16. Yogeeswaran, G. & B. S. Stein. 1980. Glycosphingolipids of metastic variant RNA virus-transformed nonproduced Balb/3T3 cells lines: Altered metabolism and cell surface exposure. J. Natl. Cancer Inst. **65**(5): 967–973.
17. Schlehofer, J. R., H. Hampl & K.-O. Habermehl. 1979. Differences in the morphology of herpes simplex virus infected cells. I. Comparative scanning and transmission electron microscopic studies on HSV-1 infected HEp-1 and chick embryo fibroblast cells. J. Gen. Virol. **44**(2): 433–442.
18. Huang, R. T. C., K. Wahn, H.-D. Klenk & R. Rott. 1980. Fusion between cell membrane and liposomes containing the glycoproteins of influenza virus. Virology **104**(2): 294–302.
19. Grey, J. E., H. J. Gitelman & A. D. Roses. 1980. Myotonic muscular dystrophy: Defective phospholipid metabolism in the erythrocyte plasma membrane. J. Clin. Invest. **65**: 1478–1482.
20. Parry, J. E., P. V. Shirodaria & C. R. Pringle. 1979. Pneumoviruses: The cell surface of lytically and persistently infected cells. J. Gen. Virol. **44**(2): 479–491.
21. Kuroda, A., T. Maeda & S.-I. Ohnishi. 1980. Enhancement of phospholipid transfer from Sendai virus to erythrocytes is mediated by target cell membrane. Proc. Natl. Acad. Sci. USA **77**(2): 804–807.
22. Barakat, H. A., D. R. Johnson & D. S. Kerr. 1980. Changes in the phospholipid composition of microsomal membranes of dystrophic hamsters. Proc. Soc. Exp. Biol. Med. **163**: 167–170.
23. Livingstone, C. J. & D. Schachter. 1980. Calcium modulates the lipid dynamics of rat hepatocyte plasma membranes by direct and indirect mechanisms. Biochemistry **19**: 4823–4827.
24. Lloyd, K. O., C. Borek, C. M. Fenoglio & D. W. King, Eds. 1980. Aging, Cancer and Cell Membranes. Thieme-Stratton, Inc. New York. pp. 282–299.
25. Andreoli, T. E., J. F. Hoffman & D. D. Fanestil, Eds. 1978. Physiology of Membrane Disorders. Plenum Press. New York.
26. Altona, J. C. & A. Vanderlaarse. 1982. Anoxia-induced changes in composition and

permeability of sarcolemmal membranes in rat heart cell cultures. Cardiovasc. Res. **16**: 138–143.
27. ROUSLIN, W., J. MACGEE, A. R. WESSELMAN, R. J. ADAMS & S. GUPT. 1980. Canine myocardial ischemia: Increased mitochondrial cholesterol, a marker of mitochondrial membrane injury. J. Molec. Cell. Cardio. **12**: 1475–1482.
28. VASDEV, S. C., K. J. KAKO & G. P. BIRO. 1979. Phospholipid composition of cardiac mitochondria and lysosomes in experimental myocardial ischemia in the dog. J. Molec. Cell. Cardio. **11**: 1195–1200.
29. CHIEN, K. R., J. ABRAMS, A. SERRONI, J. T. MARTIN & J. L. FARBER. 1978. Accelerated phospholipid degradation and associated membrane dysfunction in irreversible, ischemic liver injury. J. Biol. Chem. **253**(13): 4809–4817.
30. VASDEV, S. C., G. P. BIRO, R. NARABAITZ & K. J. KAKO. 1980. Membrane changes induced by early myocardial ischemia in the dog. Can. J. Biochem. **58**: 1112–1119.
31. JACKSON, R. J. & A. M. GOTTO, JR. 1976. Hypothesis concerning membrane structure, cholesterol and atherosclerosis. Atherosclerosis Rev. **1**: 1–21.
32. CHIEN, K. R., R. G. PFAU & J. L. FARBER. 1980. Ischemic myocardial cell injury. Am. J. Pathol. **97**(3): 505–523.
33. ROSS, R. & J. A. GLOMSET. 1973. Atherosclerosis and the arterial smooth muscle cell. Proliferation of smooth muscle is a key event in the genesis of the lesions of atherosclerosis. Science **180**: 1332–1339.
34. GUTH, P. H. 1982. Pathogenesis of gastric mucosal injury. Ann. Rev. Med. **33**: 183–186.
35. BRADY, R. O. 1982. Inherited metabolic storage disorders. Ann. Rev. Neurosci. **5**: 33–56.
36. DUNNILL, M. S. 1976. Pathological Basis of Renal Disease. Saunders. Philadelphia.
37. STIER, A. 1978. Membrane Fluidity In Biochemical Mechanisms of Liver Injury. T. F. Slater, Ed.: 319–364. Academic Press. New York.
38. BROWNLEE, M. & A. CERAMI. 1981. The biochemistry of the complications of diabetes mellitus. Ann. Rev. Biochem. **50**: 385–432.
39. PAPAHADJOPOULOS, D. 1974. Cholesterol and cell membrane function. A hypothesis concerning the etiology of atherosclerosis. J. Theoret. Biol. **43**: 329–337.
40. RODAN, S. B., R. L. HINTZ, R. I. SHA'AFI & G. A. RODAN. 1974. The activity of membrane-bound enzymes in muscular dystrophic chicks. Nature **252**: 589–590.
41. JOHANNSSON, A., C. A. KEIGHTLEY, G. A. SMITH, C. D. RICHARDS, T. R. HESKETH & J. C. METCALFE. 1981. The effect of bilayer thickness and *n*-alkanes on the activity of the (Ca^{2+} + Mg^{2+})-dependent ATPase of sarcoplasmic reticulum. J. Biol. Chem. **256**(4): 1643–1650.
42. KIMELBERG, H. K. 1977. The influence of membrane fluidity on the activity of membrane-bound enzymes. *In* Dynamic Aspects of Cell Surface Organization. G. Poste & G. L. Nicolson, Eds. **3**: 205–293. North-Holland Publishing Company. New York.
43. SANDERMANN, H., JR. 1978. Regulation of membrane enzymes by lipids. Biochim. Biophys. Acta **515**: 209–237.
44. OHLROGGE, J. B., E. A. EMKEN & R. M. GULLEY. 1981. Human tissue lipids: Occurrence of fatty acid isomers from dietary hydrogenated oils. J. Lipid Res. **22**: 955–960.
45. OHLROGGE, J. B., R. M. GULLEY & E. A. EMKEN. 1982. Occurrence of octadecanoic fatty acid isomers from hydrogenated fats in human tissue lipid classes. Lipids **17**: 511–557.
46. SCHOLFIELD, C. R. 1979. Analysis and physical properties of isomeric fatty acids. *In* Geometric and Positional Fatty Acid Isomers. E. A. Emken & H. J. Dutton, Eds. **1**: 17–52. Am. Oil Chem. Soc. Champaign, Ill.
47. LANDS, W. E. M. 1965. Effects of double bonds configuration on lecithin synthesis. J. Am. Oil Chem. Soc. **42**: 465–472.
48. LANDS, W. E. M., M. L. BLAND, L. J. NUTTER & O. S. PRIVETT. 1966. A comparison of acyl transferase *in vitro* with the distribution of fatty acids in lecithin and triglycerides *in vivo*. Lipids **1**: 224–229.
49. KINSELLA, J. B., G. BRUCHNER, J. MAI & J. SHIMP. 1981. Metabolism of trans fatty acids with emphasis on the effects of trans trans octadecadienate on lipid composition, essential fatty acids, and prostaglandins: An overview. Am. J. Clin. Nutr. **34**: 2307–2318.
50. PRIVETT, O. & M. L. BLANK. 1964. Studies on the metabolism of linoelaidic acid in the essential fatty acid-deficient rat. J. Am. Oil Chem. Soc. **41**: 292–297.

51. Demel, R. A. & B. DeKruyff. 1976. The function of sterols in membranes. Biochim. Biophys. Acta **457**: 109–132.
52. Green, C. 1977. Int. Rev. Biochem. **14**: 101–152.
53. Singer, S. & G. Nicholson. 1972. The fluid mosaic model of the structure of cell membranes. Science **175**: 720–731.
54. Kummerow, F. A. 1979. Effects of isomeric fats on animal tissue, lipid classes, and atherosclerosis. *In* Geometrical and Positional Fatty Acid Isomers. E. A. Emken & H. J. Dutton, Eds.: 151–179. Am. Oil Chem. Soc., Champaign, Ill.
55. Bird, R. P., R. K. Basrur & J. C. Alexander. 1981. Cytotoxicity of thermally oxidized fats. In Vitro **17**: 397–404.
56. Beare-Rogers, J. L. 1979. Partially hydrogenated rapeseed and marine oils. *In* Geometric and Positional Fatty Acid Isomers. E. A. Emken & H. J. Dutton, Eds.: 131–149. Am. Oil Chem. Soc., Champaign, Ill.
57. Kramsch, D. M., A. J. Aspen & L. J. Rozler. 1981. Atherosclerosis: Prevention by agents not affecting abnormal levels of blood lipids. Science **213**: 1511–1512.
58. Kramsch, D. M., A. J. Aspen & C. S. Apstein. 1980. Suppression of experimental atherosclerosis by the Ca^{2+}-antagonist lanthanum. Possible role of calcium in atherogenesis. J. Clin. Invest. **65**: 967–981.
59. Henry, P. D. & K. I. Bentley. 1981. Suppression of atherogenesis in cholesterol-fed rabbit treated with nifedipine. J. Clin. Invest. **68**: 1366–1369.
60. Katz, A. M. & F. C. Messineo. 1981. Lipid membrane interactions and the pathogenesis of ischemic damage in the myocardium. Circ. Res. **48**: 1–16.
61. Walleng, R. W. & W. E. M. Lands. 1975. Requirements for unsaturated fatty acids for the induction of respiration in saccharomyces cerevisiae. J. Biol. Chem. **250**: 9130–-9136.
62. Barnes, D. & G. Sato. 1980. Serum-free cell culture: A unifying approach. Cell **22**: 649–655.
63. Walker, B. L. 1975. Nutritional aspects of lipid research. *In* Analysis of Lipids and Lipoproteins. E. G. Perkins, Ed.: 272–284. Am. Oil Chem. Soc. Champaign, Ill.
64. Guarnieri, M. & R. M. Johnson. 1970. The essential fatty acids. Adv. Lipid Res. **8**: 115–174.
65. McMurchie, E. J. & J. K. Raison. 1976. Membrane lipid fluidity and its effect on the activation energy of membrane-associated enzymes. Biochim. Biophys. Acta **554**: 364–374.
66. Hsu, C. M. L. & F. A. Kummerow. 1977. Influence of elaidate and erucate on heart mitochondria. Lipids **12**: 486–494.
67. Decker, W. J. & W. Mertz. 1967. Effects of dietary elaidic acid on membrane fraction in rat mitochondria and erythrocytes. J. Nutr. **91**: 324–330.
68. Spector, A. A., T. L. Kaduce & R. W. Dane. 1980. Effect of dietary fat saturation on acylcoenzyme A: Cholesterol acyltransferase activity of rat liver microsomes. J. Lipid Res. **21**: 169–179.
69. Kaduce, T. L., A. B. Awad, J. J. Fontenelle & A. A. Spector. 1977. Effect of fatty acid saturation on α-aminoisobutyric acid transport in Ehrlich ascites cells. J. Biol. Chem. **252**: 6624–6630.
70. Innis, S. M. & M. T. Clandinin. 1981. Dynamic modulation of mitochondrial membrane physical properties and ATPase activity in diet lipid. Biochem. J. **198**: 167–175.
71. Galo, M. G., B. Bloj & R. N. Farias. 1975. Kinetic changes of the erythrocyte (Mg^{2+} + Ca^{2+})-adenosine triphosphatase of rats fed different fat-supplemented diet. J. Biol Chem. **250**: 6204–6207.
72. Kummerow, F. A., T. Mizuguchi, T. Arima, B. H. S. Cho & W. Y. Huang. 1978. The influence of three sources of dietary fats and cholesterol on lipid composition of swine serum lipids and aorta tissue. Artery **4**: 360–384.
73. Babka, B. 1982. The effect of different dietary fats on the lipid composition of swine cardiac plasma membrane. Ph.D. Thesis. University of Illinois. Urbana-Champaign, Ill.
74. Imai, H., N. T. Werthessen, C. B. Taylor & K. T. Lee. 1976. Angiotoxicity and arteriosclerosis due to contaminants of USP-grade cholesterol. Arch. Path. Lab. Med. **100**: 565–570.

75. Taylor, C. B., S. K. Peng, N. T. Werthessen, P. Tham & K. T. Lee. 1979. Spontaneously occurring angiotoxic derivatives of cholesterol. Am. J. Clin. Nutr. **32**: 40–47.
76. Peng, S. K., C. B. Taylor, P. Tham, N. T. Werthessen & B. Mikkelson. 1978. Effect of auto-oxidation products from cholesterol on aortic smooth muscle cells. Arch. Path. Lab. Med. **102**: 57–63.
77. Seelig, M. S. 1969. Vitamin D and cardiovascular renal and brain damage in infancy and childhood. Ann. N.Y. Acad. Sci. **147**: 537–582.
78. Toda, T., D. Leszczynski & F. A. Kummerow. 1982. Angiotoxic effects of dietary 7-ketocholesterol in chick aorta. Arter. Wall **7**: 167–175.
79. Breslow, J. L., D. A. Lothrop, D. R. Spaulding & A. A. Kandutsch. 1975. Cholesterol, 7-ketocholesterol and 25-hydroxycholesterol uptake studies and effect on 3-hydroxy-3-methyl-glutaryl-coenzyme A reductase activity in human fibroblasts. Biochim. Biophys. Acta **398**: 10–17.
80. Davie, M. & D. E. M. Lawson. 1980. Assessment of plasma 25-hydroxy vitamin D response to ultraviolet irradiation over a controlled area in young and elderly subjects. Clin. Sci. **58**: 235–242.
81. Brown, M. S. & J. L. Goldstein. 1974. Suppression of 3-hydroxy-3-methyl-glutaryl coenzyme A reductase activity and inhibition of growth of human fibroblasts by 7-ketocholesterol. J. Biol. Chem. **249**: 7306–7317.
82. Fontaine, O., T. Matsumoto, D. B. P. Goodman & H. Rasmussen. 1981. Liponomic control of Ca^{2+} transport: Relationship to mechanism of action on 1,25-dihydroxy-vitamin D_3. Proc. Natl. Acad. Sci. USA **78**: 1751–1754.
83. Rasmussen, H., T. Matsumoto, O. Fontaine & D. B. P. Goodman. 1982. Role of changes in membrane lipid structure in the action of 1,25-dihydroxyvitamin D_3. Fed. Proc. **41**: 72–77.
84. Matsumoto, T., O. Fontaine & H. Rasmussen. 1981. Effect of 1,25-dihydroxyvitamin D_3 on phospholipid metabolism in chick duodenal mucosal cell. Relationship to its mechanism of action. J. Biol. Chem. **256**: 3354–3360.
85. Christakos, S. & A. W. Norman. 1979. Studies on the mode of action of calciferol. XVIII. Evidence for a specific high affinity binding protein for 1,25-dihydroxyvitamin D_3 in chick kidney and pancreas. Biochem. Biophys. Res. Commun. **89**: 56–63.
86. Simpson, R. U. & H. F. DeLuca. 1980. Characterization of a receptor-like protein for 1,25-dihydroxyvitamin D_3 in rat skin. Proc. Natl. Acad. Sci. USA **77**: 5822–5826.
87. Hughes, M. R. & M. R. Haussler. 1978. 1,25-Dihydroxyvitamin D_3 receptors in parathyroid glands. Preliminary characterization of cytoplasmic and nuclear binding components. J. Biol. Chem. **253**: 1065–1073.
88. Coty, W. A. 1980. A specific, high affinity binding protein for 1α,25-dihydroxyvitamin D in the chick oviduct shell gland. Biochem. Biophys. Res. Commun. **93**: 285–292.
89. Ranganathan, S., J. A. K. Harmony & R. L. Jackson. 1982. Effect of Ca^{2+} blocking agents on the metabolism of low density lipoproteins in human skin fibroblasts. Biochem. Biophys. Res. Commun. **107**(1): 217–224.
90. Benga, G., A. Hodârnău, M. Ionescu, V. Pop, P. T. Frangopol, V. Strujan, R. P. Holmes & F. A. Kummerow. 1983. A comparison of the effects of cholesterol and 25-hydroxycholesterol on egg yolk lecithin liposomes: Spin-label studies. Ann. N.Y. Acad. Sci. (This volume.)
91. Holmes, R. P., M. Mahfouz, B. D. Travis, N. L. Yoss & M. J. Keenan. 1983. The effect of membrane lipid composition on the permeability of membranes to Ca^{2+}. Ann. N.Y. Acad. Sci. (This volume.)

THE EFFECT OF MEMBRANE LIPID COMPOSITION ON THE PERMEABILITY OF MEMBRANES TO Ca^{2+}*

Ross P. Holmes, Mohamedain Mahfouz, Ben D. Travis, Norma L. Yoss, and Michael J. Keenan

Department of Food Science
Burnsides Research Laboratory
University of Illinois
Urbana, Illinois 61801

Mechanisms Controlling Ca^{2+} Permeation Across Membranes

An important property of membranes is their impermeability to Ca^{2+} and other cations. This has most clearly been demonstrated in studies with phosphatidylcholine liposomes.[1] Ca^{2+}, however, is an essential cellular component that is required for many cellular functions. Furthermore, fluctuations in cytoplasmic levels of Ca^{2+} are believed to regulate many intracellular processes.[2] Mechanisms must therefore exist for Ca^{2+} to cross cellular membranes to serve its regulatory function and to meet growth requirements. Cellular mechanisms known to exist are illustrated in Figure 1. (A) denotes active transport systems, which have been identified in mitochondria, the endoplasmic reticulum, and plasma membranes. The requirement for energy to transport Ca^{2+} is clearly evident for the plasma membrane system. A concentration gradient of 10^4 across this membrane coupled with a potential difference of approximately -50 mV would favor a massive Ca^{2+} influx if Ca^{2+} was able to freely distribute across the membrane. Similar systems may exist in the nuclear envelope and in intranuclear vesicles[3] where they have been proposed to regulate Ca^{2+} fluxes during mitosis. Mechanisms must exist for Ca^{2+} to be released from the intracellular organelles that sequester Ca^{2+}, but they have not been included in Figure 1 as their nature is uncertain. (E) represents Na^+–Ca^{2+} exchange that has been identified in the plasma membranes of excitable cells but not as yet in non-excitable cells.[4] Its stoichiometry is uncertain and it is believed to act to extrude Ca^{2+} from the cell, although *in vitro* it can be manipulated to operate in either direction. (C) is a Ca^{2+} channel that probably exists in most plasma membranes to allow Ca^{2+} to selectively enter cells. This process has been most extensively studied in excitable cells where membrane depolarization results in a Ca^{2+} influx.[4] A channel sensitive to high concentrations of Verapamil has recently been shown to exist in cultured kidney cells[5] and erythrocytes.[6] (A), (C), and (E) are all protein-facilitated fluxes in contrast to (D), which represents diffusion of Ca^{2+} across lipid regions of the membrane. Interest in this mechanism has been triggered by recent observations that phosphatidic acid has Ca^{2+}-ionophoretic properties[7] and can greatly enhance the permeability of liposomes to Ca^{2+}.[1] Increased membrane synthesis of phosphatidic acid has been correlated with increased Ca^{2+} influxes in several experimental systems.[8,9] We have been unable to verify, however, that phosphatidic acid can translocate Ca^{2+} across phosphatidylcholine membranes.[10]

Physiological Significance

In an unstimulated cell, the pathways controlling Ca^{2+} fluxes are balanced to

* Supported by the University of Illinois Foundation Burnsides Research Fund.

0077-8923/83/0414-0044$01.75/0 © 1983, NYAS

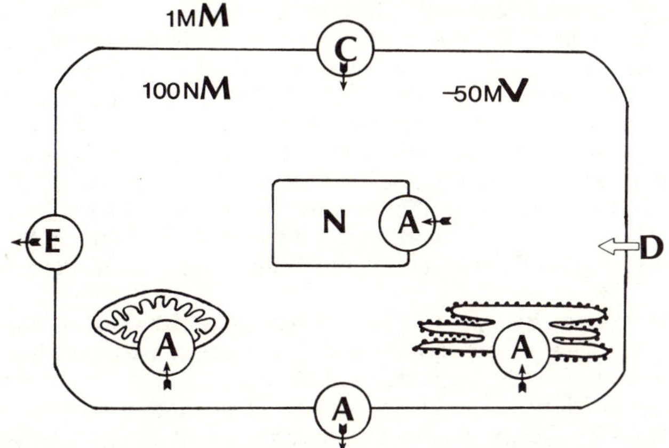

FIGURE 1. Mechanisms of Ca^{2+} flux across membranes in a typical cell. (A) denotes an active transport process; (E), a Na^{+}-Ca^{2+} exchange process; (C), a Ca^{2+}-selective channel; and (D) a diffusional process across the lipid bilayer.

maintain cytosolic Ca^{2+} at approximately 100 nM.[4,11] Following a stimulus to the cell where Ca^{2+} acts as a secondary signal, cystosolic Ca^{2+} levels rise substantially. This may be achieved by release from intracellular stores (mitochondria and endoplasmic reticulum) or by an influx from the extracellular medium. Such a signal, for example, may result either from the interaction of cell surface receptors with hormones[4] or mitogens,[12] from membrane depolarization caused by a nerve impulse in muscle contraction,[13] or from an intracellular signal (such as has been shown to occur during mitosis in an acellular slime mold).[14]

With an altered rate of influx or efflux across the plasma membrane, the potential exists to overload the intracellular sequestering systems and raise cytosolic Ca^{2+} levels to create a constant activated state. The consequences of such an overload are swelling of the endoplasmic reticulum, swelling and functional impairment of mitochondria, membrane damage resulting from the release of fatty acids by Ca^{2+}-stimulated phospholipases, and the perturbation of intracellular metabolism.[15] The final result may be cell necrosis. In cultured hepatocytes exposed to a variety of cellular toxins, which act in different ways, an enhanced influx of Ca^{2+} was a common factor in their response to the toxins.[16] Cell senescence may also be an example of the result of subtle changes in Ca^{2+} fluxes. It was recently observed that there was a correlation between senescence in cultured cells and an increased cellular Ca^{2+} content.[17]

The potential for an increased Ca^{2+} influx or decreased Ca^{2+} efflux to manifest itself in a diseased state is consequently quite apparent. In our laboratory we are investigating whether altering membrane lipid composition is a contributing factor in altering Ca^{2+} fluxes across membranes. We have been examining the effect of the incorporation of 25-hydroxylated sterols in membranes and the effect of altering the saturation and isomerization of membrane fatty acids. The 25-hydroxylated sterols we have examined are 25-hydroxycholesterol and 25-hydroxyvitamin D_3, which is actually a seco-steroid due to its broken *B* ring.

25-Hydroxycholesterol is one of the major auto-oxidation products of cholesterol and has a potent necrotic effect on cells.[18,19] It exerts an inhibitory effect on cholesterol biosynthesis and it has been suggested that its pleiotropic effects on membrane functions, including an alteration in ion permeability, stem from a reduction in membrane cholesterol content.[20] Changes in membrane cholesterol content as a contributing factor to changes in membrane permeability, however, are not supported by experiments where membrane cholesterol was altered by different means.[21] An alternative hypothesis is that the effect of 25-hydroxycholesterol on membrane permeability is related to the insertion of 25-hydroxycholesterol into membranes. Such a hypothesis would explain some of the alterations observed in Rb^+ transport in 25-hydroxycholesterol-treated Chinese hampster ovary (CHO) cells.[22] The failure of many other oxidized derivatives of cholesterol to inhibit DNA synthesis whilst still inhibiting sterol synthesis[23] does not support a theory that centers on the pleiotropic effects of 25-hydroxycholesterol being entirely caused by reduced cholesterol biosynthesis.

The other 25-hydroxylated sterol we are studying, 25-hydroxyvitamin D_3, is relevant in that, during severe hypervitaminosis D in animals, including humans, serum 25-hydroxyvitamin D levels rise more than 50-fold.[24,25] This is accompanied by severe soft tissue calcification along with substantial increases in tissue concentrations of vitamin D and 25-hydroxyvitamin D.[25] We are testing the hypothesis that the increased tissue content of Ca^{2+} is related to the insertion of 25-hydroxyvitamin D into membranes, which causes an increase in membrane permeability to Ca^{2+} and an accumulation of Ca^{2+} within cells.

We have examined the effect of fatty acid saturation on membrane Ca^{2+} permeability both *in vivo* and *in vitro*. Ever since the observations that membranes may undergo discrete phase transitions that depend on factors such as fatty acid composition, a role for a varied fatty acid composition in perturbing membrane function has been proposed.[26] An obvious extension of this proposal is that such perturbations could be contributing factors in leading to a diseased state. The role of fatty acids is apparently related to their effect on the "fluidity" properties of membranes as there is little evidence that proteins have a specific requirement for a particular species of fatty acid in order to exert their function. In recent years with the development of the fluid-mosaic model of membrane structure,[27] it has been suggested that a fluid membrane is required for optimal protein function.[26] This is supported by lower energies of activation for enzymes at temperatures above those where presumed lipid phase-transitions occur in membranes.[26] Thus, proteins can apparently undergo a conformational change more easily to catalyze a reaction in a fluid membrane than within an environment of gel-like lipids.

We have modified the fatty acid composition of erythrocyte membranes *in vivo* by altering the composition of dietary fats and have examined how these changes affect the membrane's ability to pump Ca^{2+} and Na^+ out of the cell. We have examined these active transport systems through their expression as $(Ca^{2+} + Mg^{2+})$ATPase and $(Na^+ + K^+)$ATPase activities. These enzymes appear to be ubiquitous components of plasma membranes.[28,29] The erythrocyte is a convenient model to study the response of these enzymes to membrane-lipid changes, because of the ease of obtaining source material, the ability to obtain a plasma-membrane preparation free of other contaminating subcellular membranes, and the recognized alteration in lipid composition in response to dietary lipid changes. Changes in the activity of the Ca^{2+}-pumping system in erythrocytes can initiate morphological changes[30,31] and produce cation imbalances.[31]

To determine whether the Ca^{2+} channel allowing Ca^{2+} to enter cells is influenced by the lipid composition of membranes, we have modified the fatty acid composition of an established line of kidney cells that can be grown in serum-free medium. Very little is known about the properties of this Ca^{2+} entry process; its sensitivity to the Ca^{2+} antagonist, Verapamil,[32] at high concentrations suggests that it may share some similarities with the Ca^{2+} channel in excitable cells. It is also possible that this entry process is similar to the permease that controls Ca^{2+} influx in intestinal mucosal cells in response to vitamin D. This permease is highly sensitive to the fluidity of its lipid environment,[33] responding to 1,25-dihydroxyvitamin D by increasing the phosphatidylcholine content and fatty acid unsaturation of brush border membranes, thus stimulating Ca^{2+} influx.[34]

Permeability Through the Lipid Phase: Effect of 25-Hydroxylated Sterols

Either 25-hydroxyvitamin D_3 or 25-hydroxycholesterol at 10 mole % substantially increased the permeability of phosphatidylcholine liposomes to Ca^{2+}, whereas with vitamin D_3 and cholesterol at 10 mole % the liposomes were impermeable (Figure 2). The molecular basis claimed to support observations that phosphatidic acid stimulates Ca^{2+} translocation across cellular [37] and artificial membranes[1,38] is that it acts as an ionophore. Phosphatidic acid has been shown to have ionophoretic properties in a Pressman cell[7] and to form a complex with Ca^{2+} in an organic solvent.[39] We have found that 25-hydroxyvitamin D_3 does not share these characteristics, its rate of translocation of Ca^{2+} through an organic solvent being approximately 10% of that of phosphatidic acid. Uptake by liposomes did not vary markedly with changes in phospholipid head-group but decreased with increasing saturation of phospholipid fatty acyl chains; for example, dielaidylphosphatidylcholine vesicles were only 40% as permeable to Ca^{2+} as egg yolk phosphatidylcholine vesicles. The addition of cholesterol produced a similar type of effect, indicating that the motional freedom of the phospholipid acyl chains interacting with the 25-hydroxysterols is an important factor in influencing the permeability of liposomes containing 25-hydroxylated sterols to Ca^{2+}. While the concentration of 25-hydroxysterols required to significantly

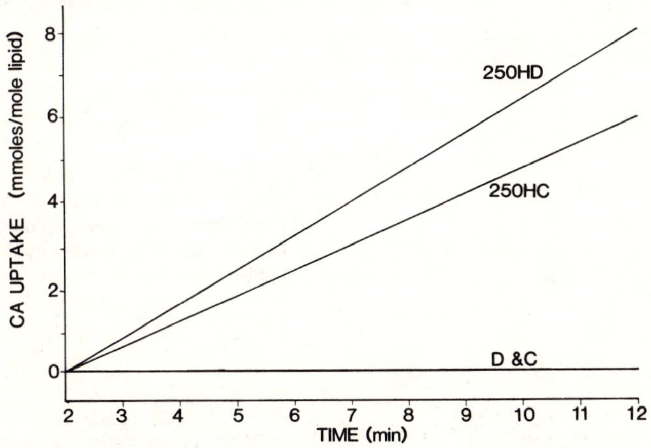

Figure 2. The effect of 25-hydroxylated sterols on the permeability of liposomes to Ca^{2+}. Liposomes (MLV) consisting of 70 mole % egg yolk PC, 20 mole % dicetyl phosphate, and 10 mole % of the indicated sterols were prepared by pressure filtration,[35] encapsulating the Ca^{2+}-sensitive dye, arsenazo III.[36] Free arsenazo III was separated from liposome-entrapped dye by Sephadex G-50 chromatography. Two to twelve minutes after the addition of Ca^{2+} to give a final concentration of 1 mM, Ca^{2+} uptake was monitored at 650 nm using a calculated molar extinction coefficient of 1.89×10^4 for the Ca^{2+}-arsenazo III complex.

alter Ca^{2+} permeability in liposomes is high at 1 mole %, it must be kept in mind that in biological membranes discrete lipid domains containing high concentrations could feasibly exist and their varied protein and lipid composition may substantially increase its effects.

To test our hypothesis that incorporation of 25-hydroxyvitamin D_3 will enhance Ca^{2+} uptake by cells, we incubated cultured kidney cells in medium containing 1 μg of 25-hydroxyvitamin D_3/ml. This concentration can be exceeded during hypervitaminosis D.[24,25] Cells were grown in serum-free medium so that vitamin D metabolites and proteins present in serum would not complicate the interpretation of the results. The 25-hydroxyvitamin D_3 was complexed to purified vitamin D binding protein (DBP) to simulate physiological conditions. From other experiments we have done on 25-hydroxyvitamin D_3 incorporation by these cells, we estimate that in this medium 20–50 pmoles 25-hydroxyvitamin D_3 were incorporated per mg cellular protein. When Ca^{2+} uptake was compared in these cells with cells grown in medium containing binding protein alone, there was no observable difference (FIGURE 3). Before discounting the hypothesis that cellular incorporation of 25-hydroxyvitamin D_3 increases their permeability to Ca^{2+}, several factors have to be considered. First, recent results of Varecka and Carafoli[6] indicate that, in the erythrocyte, substantial cycling of Ca^{2+} across the plasma membrane occurs; that is, the majority of Ca^{2+} that flows into the cell is immediately pumped out again. If a similar process operates in the kidney cells used in this study, an increased Ca^{2+} uptake may be masked by an increased efflux of Ca^{2+}. If this does occur, it would suggest that while an increased permeability results, accumulation of Ca^{2+} above normal does not occur. Second, a more precise way to determine whether Ca^{2+} accumulation occurs with exposure to 25-hydroxyvitamin D_3 would be to determine if a change in cellular Ca^{2+} content occurred. Third, it is possible that other serum components are required for increased Ca^{2+} accumulation to occur. We are currently testing this using serum depleted of endogenous binding protein and vitamin D by passage through a binding protein antibody-affinity column. Fourth, the kidney cells used may not be susceptible to the ac-

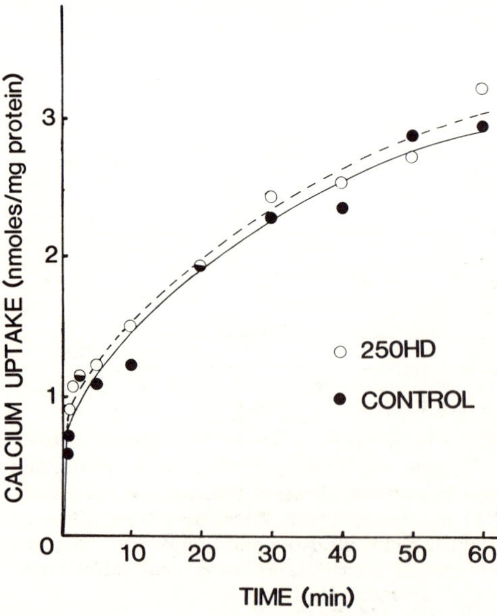

FIGURE 3. The effect of 25-hydroxy vitamin D_3 on Ca^{2+} uptake by MDCK cells. Confluent monolayers of cells maintained in a hormone-supplemented, serum-free medium[40,41] were incubated for 24 hours at 37°C in medium containing per ml either 1 μg 25-hydroxyvitamin D_3 and 400 μg purified DBP (○) or 400 μg DBP alone (●). To determine Ca^{2+} uptake, the medium was removed, replaced with fresh medium containing $^{45}CaCl_2$ (1 μCi/ml) and incubated at 37°C for the indicated time. Cells were washed five times with ice-cold buffer consisting of 150 mM NaCl, 5 mM $CaCl_2$, 0.1 mM $LaCl_3$, and 10 mM HEPES pH 7.4.[37]

tion of 25-hydroxyvitamin D_3, particularly as we have recently observed that they retain their *in vivo* capacity to metabolize 25-hydroxyvitamin D_3 to 24,25-dihydroxyvitamin D_3 and, under the conditions of this experiment, may convert a large amount of the 25-hydroxyvitamin D_3 to 24,25-dihydroxyvitamin D_3. Smooth muscle cells, for instance, may be more sensitive to 25-hydroxyvitamin D_3, as their necrosis is apparent during hypervitaminosis D[42] and they are not known to be target cells responsive to vitamin D or its metabolites.

If 25-hydroxyvitamin D_3 is shown to enhance Ca^{2+} accumulation in cells, the next step to determine the validity of our hypothesis is to show that some 25-hydroxyvitamin D_3 is incorporated into the plasma membrane.

Effect of Membrane Fatty Acid Composition on Ca^{2+} Uptake by MDCK Cells

The effect of supplementing the medium with either 18:0 (stearic acid) or 18:2 (linoleic acid) complexed to bovine serum albumin (BSA) on the total fatty acid content of MDCK cells grown in serum-free medium is shown in Table 1. Substantial differences in saturated, monounsaturated, and polyunsaturated fatty acid contents are apparent. Notably, the cells grown on 18:0 responded by synthesizing a large amount of oleic acid (18:1). This suggests that Δ9 desaturase activity increased to maintain a required level of fluidity in cellular membranes. Cells grown on 18:2 produced a range of more polyunsaturated acids apparently derived from 18:2. Cells grown on 18:0 produced fewer polyunsaturated acids. These included derivatives of 18:1, derivatives of 18:2 arising from trace contaminants in the BSA, and some apparent ω7 derivatives of 16:1. The presence of this latter family of fatty acids is based

Table 1

Modification of MDCK Fatty Acid Composition by Media Supplementation

	% of Total Fatty Acids*	
Fatty Acid	18:0	18:2
16:0	11.4	9.7
16:1ω7	8.5	3.1
18:0	11.9	8.1
18:1ω9	39.3	10.0
18:2†	2.7	38.0
18:3†	1.0	4.0
20:1ω9	3.8	–
20:2†	2.7	0.7
20:3†	2.9	12.3
20:3ω9	4.7	2.6
22:1ω9	2.1	–
20:4†	1.9	4.0
24:0	1.4	0.5
24:1	2.3	–
22:5	–	0.9
Others	3.4	6.1

* Fatty acid methyl esters were analyzed on a 60 m SP2340 glass capillary column in a Packard 428 gas chromatograph using a temperature program increasing the temperature from 160°C to 220°C at 1°/min.

† These results represent mixtures of ω6 and ω7 isomers.

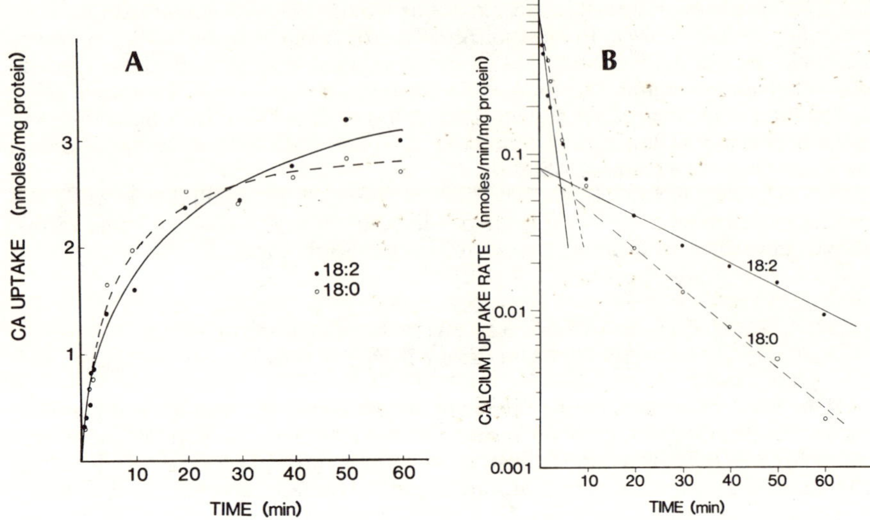

FIGURE 4. The effect of fatty acid supplementation on Ca^{2+} uptake by MDCK cells. (A) Ca^{2+} uptake was determined as described in the legend to FIGURE 3. In (B), a linear-regression analysis of the data yielded for the rapidly exchanging compartment a flux rate (J, in nmoles/mg protein/min) of 0.67 and a compartment size (S, in nmoles/mg protein) of 1.37 for 18:2 cells, whereas for 18:0 cells the values were 0.67 and 2.15, respectively. The values for the slow influx into the cells were $J = 0.092$ and $S = 2.37$ for 18:2 grown cells, and $J = 0.085$ and $S = 1.41$ for 18:0 grown cells.

on the fatty acid composition of cells grown in medium without BSA or fatty acid supplements, and the known existence of this minor pathway in mammalian cells.[43] Similar changes that have been observed in the phospholipids of LM[44,45] and CHO[46] cells have been correlated with changes in membrane fluidity. Supplementing cells with saturated fatty acids produced less fluid membranes than when supplemented with unsaturated fatty acids. In LM cells, a reduction in adenylate cyclase activity[47] occurred in concert with a reduction in fluidity.[48] Ca^{2+} uptake by MDCK cells grown in 18:0 and 18:2-containing media is illustrated in FIGURE 4A. The difference between these two uptake curves is more clearly observed when the exponential change in the rate of uptake is analyzed as recommended by Borle.[49] Such an analysis is required to discriminate between true changes in flux rates or changes in pool sizes.[50] This plot (FIGURE 4B) clearly shows that the fluxes relating Ca^{2+} binding to the cell surface and Ca^{2+} entry into intracellular pools, denoted by the intercepts on the x-axis, are similar. The difference in uptake curves is accounted for by changes in pool sizes, both the surface-binding pool and the intracellular pool. The nature of these changes in cellular components is not known. Before concluding that the bulk changes in membrane fluidity, which most likely occurred in the experiments reported here, did not affect the protein(s) facilitating Ca^{2+} entry into the kidney cells, some reservations must be made. If Ca^{2+} cycling does occur across the plasma membrane in a manner similar to that observed in erythrocytes,[41] a decreased influx through the Ca^{2+} channel could be offset by a decreased efflux through the Ca^{2+} pump. In other words, alterations in membrane fluidity affect both reactions to the same extent.

EFFECT OF ERYTHROCYTE MEMBRANE FATTY ACID CHANGES ON $(Ca^{2+} + Mg^{2+})$ATPASE AND $(Na^{+} + K^{+})$ATPASE ACTIVITIES

We have examined the effect of varying the dietary fatty acid composition on the lipid composition of rat erythrocyte membranes and its effect on cation-transporting enzymes. The activities of both $(Ca^{2+} + Mg^{2+})$ATPase[51] and $(Na^{+} + K^{+})$ATPase[52] are sensitive to the nature of their lipid environment. Semi-purified diets, prepared according to the recommendations of the American Institute of Nutrition,[53] were supplemented with 10% (w/w) of either lard (L), safflower oil (SO), corn oil (CO), hydrogenated soybean oil:corn oil mixture (85:15) (HSBO), margarine-extracted fat (M), and hydrogenated coconut oil (HCO). Diets were fed *ad libitum* to male Sprague-Dawley rats for six to eight weeks from weaning for all experimental diets except when CO and HCO were compared, where male Lewis rats were fed the diet for seven to eight weeks. The dietary fats, whose fatty acid composition is given in TABLE 2, were chosen to allow testing of the effects of fatty acid saturation by comparing L, SO, and CO; the effects of isomerization by comparing the relationship of two levels of isomeric fatty acids, HSBO and M, to L, SO, and CO; and the effects of essential fatty acid deficiency by comparing HCO and CO.

Fatty Acid Saturation

The major changes occurring in the fatty acid composition of erythrocyte membranes from rats fed L, SO, or CO were in 18:1 and 18:2 fatty acids (TABLE 3). No effect of these diets was observed on $(Na^{+} + K^{+})$ATPase activity in these membranes, nor were there any apparent trends (TABLE 4). A significant effect was observed, however, when $(Ca^{2+} + Mg^{2+})$ATPase activity was examined. The animals fed the lard diet had a basal activity that was 20% lower than that of animals fed the CO diet, and a calmodulin-stimulated activity that was 12% lower. The SO diet membranes had activities intermediate to those of L and CO, which is consistent with the effect of the saturation of dietary fatty acids on the activity of this enzyme.

TABLE 2

FATTY ACID COMPOSITION OF DIETARY FATS

	% of Total Fatty Acids					
Fatty Acid	L	SO	CO	HSBO	M	HCO
12:0	–	–	–	–	–	70
14:0	1.8	0.1	–	–	–	13.4
16:0	21.5	5.3	12.5	10.6	13.7	7.9
18:0	12.4	2.5	2.2	10.6	9.4	7.6
18:1 *trans**	–	–	–	39.3	19.0	–
18:1 *cis**	42.2	74.3	26.1	23.6	26.8	–
18:2†	11.9	16.0	57.4	14.6	24.9	–
18:3	3.2	0.4	0.8	–	2.7	–

* A mixture of positional isomers.

† These results represent *cis* Δ9,12–18:2. Other positions and geometrical isomers were separated on the SP2340 capillary column and accounted for less than 0.5% of the total fatty acids.

TABLE 3

FATTY ACID COMPOSITION OF ERYTHROCYTES

Fatty Acid	% of Total Fatty Acids					
	L	SO	CO	HSBO	M	HCO
16:0	27.9	28.5	29.6	24.7	23.4	26.1
18:0	15.6	14.7	13.5	11.1	11.7	14.8
18:1t	0.2	0.9	0.3	6.7	6.2	0.3
18:1	11.1	14.8	8.9	11.5	12.1	13.9
18:2	7.3	6.6	12.2	8.9	8.4	4.7
20:3ω6	0.4	0.4	0.5	0.6	0.6	1.2
20:3ω9	0.3	0.4	0.1	0.2	0.2	3.9
20:4	26.0	25.7	26.3	25.0	24.0	22.7
22:5	1.0	0.6	0.7	0.5	1.3	0.7
22:6	2.2	1.5	1.6	2.1	2.6	1.6

Fatty Acid Isomerization

The margarine stock, HSBO, which is blended with a vegetable oil to produce margarine, and the margarine-extracted fat, M, both contain a range of *cis* and *trans* isomers. The double-bond position of the octadecenoic acids appeared to be predominantly between Δ6 and Δ13, which is consistent with reports of other investigators.[56,57] Only trace amounts of *trans,trans* 18:2 ($< 0.2\%$), *cis,trans* 18:2 ($< 0.2\%$), and *trans,cis* 18:2 ($< 0.2\%$) were identified in HSBO and none in M. The most meaningful comparison to make concerning the effect of isomeric fatty acids is between HSBO, L, and SO. HSBO contains a level of 18:2 (obtained through its blending with CO) intermediate to those of L and SO. The primary difference between the fats is that saturated and monounsaturated fatty acids are partially replaced by isomeric fatty acids. As the physical properties of *trans* monounsaturated isomers (e.g. melting points) are generally intermediate to those of *cis* isomers and saturated fatty

TABLE 4

THE EFFECT OF DIETARY FAT ON (Ca + Mg)ATPASE AND ERYTHROCYTE (Na + K)ATPASE ACTIVITIES

Fat	$(Na^+ + K^+)$ATPase	$(Ca^{2+} + Mg^{2+})$ATPase	Calmodulin-stimulated $(Ca^{2+} + Mg^{2+})$ATPase
HSBO	424	509*	1928*
CO	446	492*	1886*
M	491	456	1830
SO	481	455	1772
L	418	395†	1664†

Erythrocyte ghosts were prepared essentially as described by Vincenzi and Farrance.[55] $(Ca^{2+} + Mg^{2+})$ATPase activity was measured by the difference in the inorganic phosphate release in 1 hour at 37°C following the incubation of 100–250 μg of membrane protein in 1 ml of assay medium containing 67 mM NaCl, 12.5 mM KCl, 15 mM imidazole, 15 mM histidine, 2.5 mM $MgCl_2$, 3 mM ATP, 0.1 mM ouabain, pH 7.2 in the presence and absence of 200 μM $CaCl_2$. Calmodulin stimulation was measured by the addition of 1 μg bovine calmodulin. $(Na^+ + K^+)$ ATPase activity was measured by omitting ouabain. Activities, expressed in nmoles P_i released mg protein^{-1} hr^{-1}, are the means of at least five different preparations and those with different superscripts are significantly different at the $p < 0.05$ level.

acids, the effect of *trans* isomers when incorporated into phospholipids, if purely physical, could be expected to lie between those of *cis* monounsaturated and saturated fatty acids. If the activity of (Ca^{2+} + Mg^{2+})ATPase is influenced by the physical properties of phospholipids, the activity in membranes from HSBO-fed rats should lie between those of L and SO. This was not observed, however, as their activity was similar to that in the CO membranes. The reasons for this effect are not clear. The fatty acid composition of the HSBO-fed rat erythrocytes differed from L and SO membranes primarily in that the *trans* 18:1 isomers were incorporated at the expense of saturated fatty acids. The 18:2 content was more similar to that of L and SO than to CO. The effect with the margarine diet was consistent with the effect of HSBO, although interpretation of its effects are hampered by its containing a level of 18:2 different from those of L, SO, and HSBO.

Essential Fatty Acid Deficiency

In view of the effect of fatty acid saturation discussed above, the enzyme activities in erythrocyte membranes from rats fed an essential fatty acid (EFA)–deficient diet were unexpected. Membranes from rats fed the HCO diet, which contained only saturated fatty acids, had higher (Ca^{2+} + Mg^{2+})ATPase and (Na^{+} + K^{+})ATPase activities than membranes from rats fed CO (TABLE 5). Sun and Sun[58] observed that the (Na^{+} + K^{+})ATPase activity of synaptosomes increased with EFA deficiency, which is consistent with our results. That the rats used in our study suffered from marginal EFA deficiency is evident from their decreased 18:2 content, their marginally lower 20:4 content and the appearance of 4% 20:3 ω9, the "deficiency triene." In agreement with the observations of other workers,[59,60] we have not been able to detect any changes in cholesterol-to-phospholipid ratios in these membranes (or in those of any of the other experimental groups). In similar types of experiments, other workers have found no change in phospholipid classes.[61] We also examined the properties of the spin label, 5-doxylstearic acid, when incorporated into HCO and CO membranes. No differences were found in the order parameter of the spin label in agreement with the electron spin resonance (ESR) spectral analyses of Ehrstrom *et al.*[59] $S = 0.704 \pm 0.001$ for HCO membranes and $S = 0.703 \pm 0.001$ for CO membranes. This result reflects the averaging of the spin resonance of the probe incorporated into possibly multiple lipid environments in the membrane and, in effect, shows the insensitivity of the technique when applied to the erythrocyte membrane.

There are several possible explanations for the increased enzyme activities observed with the EFA-deficient diet. (1) The differences in fatty acid chain length be-

TABLE 5

THE EFFECT OF EFA DEFICIENCY ON ERYTHROCYTE (Na + K)ATPASE AND (Ca + Mg)ATPASE ACTIVITIES

Fat	(Na^{+} + K^{+})ATPase	(Ca^{2+} + Mg^{2+})ATPase	Calmodulin-stimulated (Ca^{2+} + Mg^{2+})ATPase
CO	407*	788*	1885*
HCO	510†	1160†	2680†

Enzyme activities were measured as in TABLE 4 except that the incubation assay contained 8 mM NaCl, 15 mM KCl, and 3 mM $MgCl_2$. Activities, expressed in nmoles P_i released mg protein^{-1} hr^{-1}, are the means of at least four different preparations and those with different superscripts are significantly different at the $p < 0.01$ level.

tween the two diets produced subtle but important effects in membrane fatty acid composition. (2) There is a change in a particular molecular species of phospholipid that may specifically associate with the enzymes, e.g., a species containing 20:3 ω9 may produce a high activity. (3) The essential fatty acid deficiency may produce a general change that affects many membrane proteins, such as an alteration in protein phosphorylation or methylation. (4) There is a change in a lipid class, such as the free fatty acids of the membrane that has not yet been recognized as occurring or being of importance. (5) The protein composition of the erythrocytes is altered and increased amounts of the ATPases are present relative to other membrane proteins. In view of the observed shedding of erythrocyte antigens induced by membrane rigidification,[62] changes in protein composition might be expected.

We are currently undertaking a more in-depth analysis of the changes in enzyme kinetic properties, the number of enzymes per cell, the phospholipid composition, and the changes occurring in the fluorescence properties of probe molecules, in all experimental diets to more fully understand the changes that occurred.

Conclusions

The main objective in our laboratory has been to determine whether dietary lipids can modify membrane structure and function. In this way we may be able to demonstrate what role, if any, they play in the initiation and development of diseased states. We have sought answers to this question through an examination of the way in which dietary lipids modify membrane permeabilities to Ca^{2+}, a parameter crucial to normal cell function.

Evidence presented here indicates that membrane-lipid composition is important in determining the rate of Ca^{2+} permeation through a lipid bilayer using 25-hydroxylated sterols as an example. Clarification of the biological effects of these sterols requires further experimentation. While we were unable to show an effect of membrane fatty acid saturation and presumed membrane fluidity changes on the Ca^{2+} channel facilitating Ca^{2+} entry into cells, we were able to demonstrate that changes in the rate of Ca^{2+} efflux can be influenced *in vivo* by changes in the fatty acid composition of dietary fats. Essential fatty acid deficiency increased the activity of membrane-bound enzymes despite the dietary fat used to induce deficiency being completely saturated. This clearly demonstrates that in *in vivo* experiments the effect of the saturation of dietary fatty acids and the effect of fatty acid deficiency must be segregated.

Acknowledgments

We wish to thank Dr. G. Boissonneault and Dr. P. V. Johnston for supplying blood from the rats on HCO and CO diets referred to in Table 5. We also wish to thank Dr. P. Morse for the use of the Varian ESR spectrometer in his laboratory and for assistance with the spectral analyses. The ESR facility is supported in part by the National Foundation for Cancer Research. This work would not have been possible without the support and encouragement of Dr. F. A. Kummerow.

References

1. Serhan, C., P. Anderson, E. Goodman, P. Dunham & G. Weissmann. 1981. J. Biol. Chem. **256**: 2736–2741.

2. Rasmussen, H., P. Jensen, W. Lake & D. B. P. Goodman. 1976. Clin. Endocrin. **5**: 11S–27S.
3. Silver, R. B., R. D. Cole & W. Z. Cande. 1980. Cell **19**: 505–516.
4. Borle, A. B. 1981. Rev. Physiol. Biochem. Pharmacol. **90**: 13–153.
5. Sandvig, K. & S. Olsnes. 1982. J. Biol. Chem. **257**: 7495–7503.
6. Varecka, L. & E. Carafoli. 1982. J. Biol. Chem. **257**: 7414–7421.
7. Tyson, C. A., H. V. Zande & D. E. Green. 1976. J. Biol. Chem. **251**: 1326–1332.
8. Putney, J. W., Jr., S. J. Weiss, C. M. Van de Walle & R. A. Haddas. 1980. Nature **284**: 345–347.
9. Salmon, D. M. & T. W. Honeyman. 1980. Nature **284**: 344–345.
10. Holmes, R. P. & N. L. Yoss. 1983. Nature. (In press.)
11. Rink, T. J., R. Y. Tsien & A. E. Warner. 1980. Nature **283**: 658–660.
12. Tsien, R. Y., T. Pozzan & T. J. Rink. 1982. Nature **295**: 68–71.
13. Baker, P. F. 1976. Symp. Soc. Exp. Biol. **30**: 67–88.
14. Holmes, R. P. & P. R. Stewart. 1977. Nature **269**: 592–594.
15. Trump, B. F., I. K. Berezesky & A. R. Osornio-Vargas. 1981. *In* Cell Death in Biology and Pathology. I. D. Bowen & R. A. Lockshin, Eds.: 209–242. Chapman and Hall. New York.
16. Schanne, F. A. X., A. B. Kane, E. E. Young & J. L. Farber. 1979. Science **206**: 700–702.
17. Shapiro, B. L. & L. F.-H. Lam. 1982. Science **216**: 417–419.
18. Peng, S.-K., P. Tham, C. B. Taylor & B. Mikkelson. 1979. Am. J. Clin. Nutr. **32**: 1033–1042.
19. Imai, H., N. T. Werthessen, V. Subramanyam, P. W. LeQuesne, A. H. Soloway & M. Kanisawa. 1980. Science **207**: 651–653.
20. Kandutsch, A. A., H. W. Chen & H.-J. Heiniger. 1978. Science **201**: 498–501.
21. Bakker-Grunwald, T. & M. Sinensky. 1979. Biochim. Biophys. Acta **558**: 296–306.
22. Chen, H. W., H.-J. Heiniger & A. A. Kandutsch. 1978. J. Biol. Chem. **253**: 3180–3185.
23. Defay, R., M. E. Astruc, S. Roussillon, B. Descomps & A. Crastes de Paulet. 1982. Biochem. Biophys. Res. Commun. **106**: 362–372.
24. Shepard, R. M. & H. F. DeLuca. 1980. Arch. Biochem. Biophys. **202**: 43–53.
25. Holmes, R. P. 1982. *In* Chemical, Biochemical and Clinical Endocrinology of Calcium Metabolism. A. W. Norman, K. Schaeffer, D. V. Herrath & H. G. Grigoleit, Eds.: 567–569. Walter De Gruyter. New York.
26. Kimelberg, H. K. 1977. Cell Surface Rev. **3**: 205–293.
27. Singer, S. J. & G. L. Nicolson. 1972. Science **175**: 720–731.
28. Barritt, G. J. 1981. Trends Biochem. Sci. **6**: 322–325.
29. Evans, W. H. 1979. *In* Laboratory Techniques in Biochemistry and Molecular Biology. T. S. Work & E. Work, Eds. **7**: 1–266. North Holland, New York..
30. Feig, S. A. & G. Guidotti. 1974. Biochem. Biophys. Res. Commun. **58**: 487–494.
31. Dunn, M. J. & R. Grant. 1974. Biochim. Biophys. Acta **352**: 97–116.
32. Sandvig, K. & S. Olsnes. 1982. J. Biol. Chem. **257**: 7495–7503.
33. Fontaine, O., T. Matsumoto, D. B. P. Goodman & H. Rasmussen. 1981. Proc. Natl. Acad. Sci. USA **78**: 1751–1754.
34. Matsumoto, T., O. Fontaine & H. Rasmussen. 1981. J. Biol. Chem. **256**: 3354–3360.
35. Olson, F., C. A. Hunt, F. C. Szoka, W. J. Vail & D. Papahadjopoulos. 1979. Biochim. Biophys. Acta **557**: 9–23.
36. Weissmann, G., P. Anderson, C. Serhan, E. Samuelsson & E. Goodman. 1980. Proc. Natl. Acad. Sci. USA **77**: 1506–1510.
37. Ohsako, S. & T. Deguchi. 1981. J. Biol. Chem. **256**: 10945–10948.
38. Serhan, C. N., J. Fridovich, E. J. Goetzl, P. B. Dunham & G. Weissmann. 1982. J. Biol. Chem. **257**: 4746–4752.
39. Cullis, P. R., B. De Kruijff, M. J. Hope, R. Nayer & S. L. Schmid. 1980. Can. J. Biochem. **58**: 1091–1100.
40. Taub, M., L. Chuman, M. H. Saier, Jr. & G. Sato. 1979. Proc. Natl. Acad. Sci. USA **76**: 3338–3342.
41. Taub, M. & G. H. Sato. 1979. J. Supramol. Struc. **11**: 207–216.
42. Kamio, A., F. A. Kummerow & I. Hideshige. 1977. Arch. Pathol. Lab. Med. **101**: 378–381.

43. Sprecher, H. 1977. *In* Polyunsaturated Fatty Acids. W.-H. Kunau & R. T. Holman, Eds.: 1-18. American Oil Chemistry Society, Champaign, Ill.
44. Williams, R. E., B. J. Wisnieski, H. G. Rittenhouse & C. F. Fox. 1974. Biochemistry **13**: 1969-1977.
45. Ferguson, K. A., M. Glaser, W. H. Bayer & P. R. Vagelos. 1975. Biochemistry **14**: 146-151.
46. Rintoul, D. A., L. A. Sklar & R. D. Simoni. 1978. J. Biol. Chem. **253**: 7447-7452.
47. Engelhard, V. H., M. Glaser & D. R. Storm. 1978. Biochemistry **17**: 3191-3200.
48. Gilmore, R., N. Cohn & M. Glaser. 1979. Biochemistry **18**: 1042-1049.
49. Borle, A. B. 1969. J. Gen. Physiol. **53**: 43-56.
50. Borle, A. B. 1981. Cell Calcium **2**: 187-196.
51. Niggli, V., E. S. Adunyah & E. Carafoli. 1981. J. Biol. Chem. **256**: 8588-8592.
52. DePont, J. J. H. H. M. & S. L. Bonting. 1977. Adv. Exp. Med. Biol. **83**: 219-224.
53. Bieri, J. G., G. S. Stoewsand, G. M. Briggs, R. W. Phillips, J. C. Woodard & J. J. Knapka. 1976. J. Nutr. **106**: 1340-1348.
54. Vincenzi, F. F. & M. L. Farrance. 1977. J. Supramol. Struc. **7**: 301-306.
55. Piper, J. M. & S. J. Lovell. 1981. Anal. Biochem. **117**: 70-75.
56. Scholfield, C. R., V. L. Davison & H. J. Dutton. 1967. J. Am. Oil Soc. **44**: 648-651.
57. Carpenter, D. L. & H. T. Slover. 1973. J. Am. Oil Soc. **50**: 372-376.
58. Sun, G. Y. & A. Y. Sun. 1974. J. Neurochem. **22**: 15-18.
59. Watson, W. C. 1963. Br. J. Haematol. **9**: 32-38.
60. Ehrstrom, M., M. Harms-Ringdahl & C. Alling. 1981. Biochim. Biophys. Acta **644**: 175-182.
61. De Gier, J. & L. L. M. Van Deenen. 1964. Biochim. Biophys. Acta **84**: 294-304.
62. Muller, C. P. & M. Shinitzky. 1981. Exp. Cell Res. **136**: 53-62.

INTERACTION BETWEEN LYSOLECITHIN AND PLATELETS AND ITS RELATIONSHIP TO DISEASE

Mircea Petru Cucuianu

Clinical Chemistry and Clinical Pathology
Faculty of Medicine
Medical and Pharmaceutical Institute of Cluj-Napoca
R-3400 Cluj-Napoca, Romania

Lysolecithin (LPC) influences a number of membrane properties, including membrane fusion, cell-surface morphology, and the activity of membrane-bound enzymes.[1] Recent evidence suggests that LPC represents the main pathway for the incorporation of polyunsaturated fatty acids into platelet phospholipids for the subsequent remodeling of platelet membranes,[2] and that LPC concomitantly arises with the generation of eicosanoids.[3] Platelet functions such as aggregation, secretion, and adhesiveness are greatly dependent on surface-membrane properties. The effect of LPC on platelet properties is a question that has been addressed in our laboratory. In this article I examine the response of platelet functions to exogenous LPC and the relationship of this response to diseased states.

Effect of LPC on Platelet Aggregation

The first question to ask when considering the relationship of exogenous (plasma) LPC to platelet function is whether an exchange of LPC occurs between plasma lipoproteins and platelets. Transfer was confirmed by incubating unlabeled, washed platelets in platelet-free plasma (PRP), in which all the major phospholipids were labeled *in vivo* with $^{32}PO_4$.[4] Up to 43% of platelet LPC was rapidly labeled. Labeling of platelet phosphatidylcholine (PC), however, occurred more slowly and reached a maximum of only 7%. These results suggested that uptake and incorporation of endogenous plasma LPC by platelets may be an important mechanism in the modification of platelet-membrane phospholipid fatty acid composition and possibly platelet function.

Data concerning the effect of LPC upon platelet function are rather conflicting. It has been suggested that LPC might represent the "transferable factor," causing abnormal electrophoretic behavior in platelets isolated from patients with vascular disease.[6] This abnormal behavior, characterized by a greater increase in platelet electrophoretic mobility induced by ADP (5 ng) than by a concentration of this nucleotide ten times higher, could be reproduced with normal platelet-rich plasma–platelet-poor plasma (PRP-PPP) mixtures by adding LPC in concentrations of 10–100 μg/ml.[6] However, the exposure of platelets in citrated PRP to LPC, in a concentration similar to that found in normal plasma, inhibits irreversible platelet aggregation induced by either ADP, adrenalin, collagen, thrombin, or serotonin.[7,8] It was suggested that this inhibition was caused by the blocking of the normal platelet-release reaction and was probably linked to the surface activity of LPC. Interestingly, saturated LPC fractions inhibited aggregation, but a LPC fraction containing a high percentage of polyunsaturated fatty acids did not. Thus, in contrast to data concerning electrophoretic behavior, results related to platelet aggregation suggest that LPC may have a thromboprotective role.

0077-8923/83/0414–0057$01.75/0

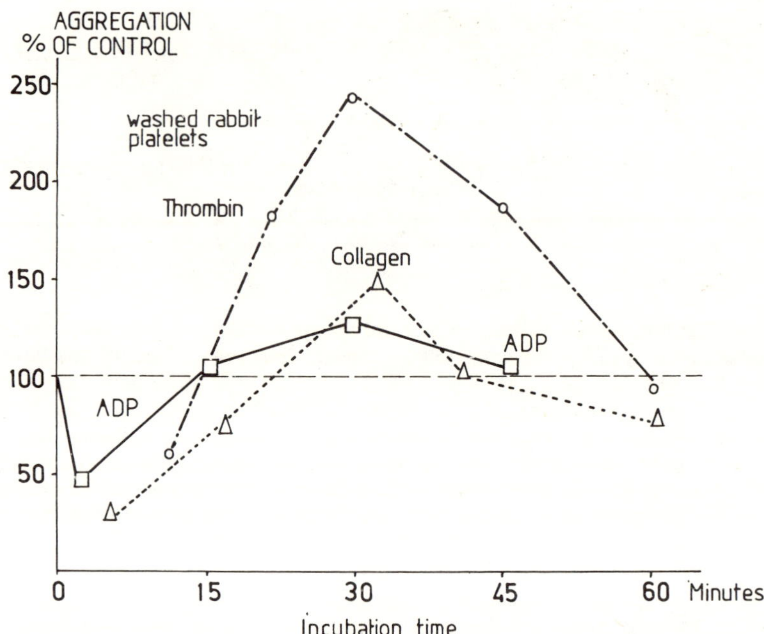

FIGURE 1. Inhibition and subsequent potentiation of platelet function by lysolecithin. LPC (50 μM in final concentration) was added to washed rabbit platelets in suspension. After various periods of incubation the platelets were stimulated by either ADP (□), thrombin (○), or collagen (△). (From Joist *et al.*[9] With permission from *Blood*).

These conflicting results led us to systematically study the behavior of platelets exposed for various time intervals to various concentrations of LPC.[9] LPC, when added to human or rabbit PRP, inhibited aggregation induced by low concentrations of ADP (3–5 μM). Inhibition was evident with final concentrations of added LPC as low as 100 μM. Similar inhibitory effects were noted for platelet aggregation induced by thrombin, collagen, and epinephrine. At concentrations of LPC below 200 μM, the inhibitory effect on aggregation induced by ADP, collagen, or thrombin in citrated PRP was partially reversible over a period of 60–90 min. The transient nature of the inhibition was evident when platelets were incubated with 50 μM LPC (FIGURE 1). It caused an initial inhibition that was followed by a subsequent potentiation of platelet aggregability and of the release reaction, particularly with thrombin. These results have been confirmed by Bekemeir *et al.*,[10] who recently showed that low concentrations (40 μM) of LPC caused a potentiation of ADP-induced platelet aggregation in rat PRP, whereas concentrations above 200 μM were inhibitory. That the physiological plasma concentrations of LPC seem to be in the range of the critical concentration between potentiation and inhibition is noteworthy. Such data suggest that LPC may play a regulatory role in platelet aggregability.

Another approach would be to study the effect of LPC upon the interactions among platelets, the plasma von Willebrand factor, and ristocetin. LPC caused a slight and inconstant potentiation of ristocetin-induced platelet aggregation (FIGURE 2). To clarify the effect that LPC has in this system, a more extensive variation of the incubation conditions is required. During this investigation a striking phenomenon

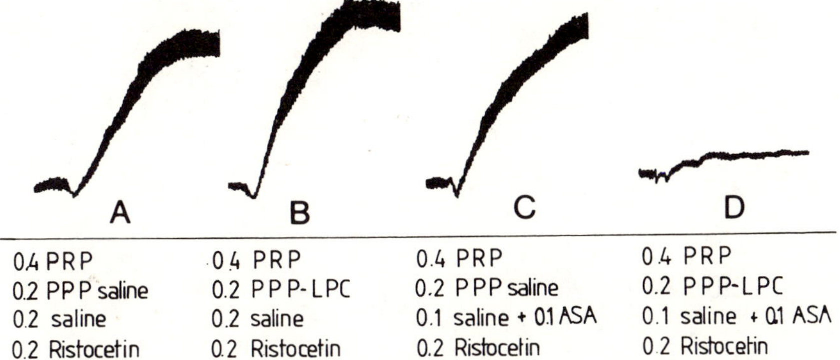

A	B	C	D
0.4 PRP	0.4 PRP	0.4 PRP	0.4 PRP
0.2 PPP saline	0.2 PPP-LPC	0.2 PPP saline	0.2 PPP-LPC
0.2 saline	0.2 saline	0.1 saline + 0.1 ASA	0.1 saline + 0.1 ASA
0.2 Ristocetin	0.2 Ristocetin	0.2 Ristocetin	0.2 Ristocetin

FIGURE 2. Influence of LPC upon ristocetin-induced platelet aggregation and its interaction with inhibitors of platelet function. In B, 0.4 ml PRP was incubated for 7 min with 0.2 ml of a mixture containing equal amounts of PPP and a solution of LPC (4 mM) in saline. The final concentration of added LPC in this incubation mixture was about three times higher than physiological concentrations (200–300 μM). After incubation, 0.2 ml saline and 0.2 ml ristocetin (5.2 mg/ml) were rapidly added, and platelet aggregation was recorded for 5 min. In A, saline was substituted for LPC. 0.1 ml of aspirin (50 mg/ml) was added in C and D. 0.1 ml of indomethacin (100 mg/ml) produced the same effects.

became apparent. In contrast to control samples, ristocetin-induced platelet aggregation in LPC-treated PRP was completely inhibited by aspirin or indomethacin. The same total inhibition was noted with 0.1 ml of a dipyridamol solution (1 mM). In the absence of LPC, these inhibitors had no effect upon ristocetin-induced platelet aggregation. A report by Jenkins *et al.*[11] indicated that various inhibitors of ADP-induced platelet aggregation, such as EDTA, EGTA, PGE, and adenosine, did not inhibit the aggregation response to ristocetin, but prevented the appearance of the second wave observed in PRP. In the same study, aspirin did not inhibit the initial response of ristocetin-induced platelet aggregation but depressed the second wave caused by the release reaction. In our experiments there was no relationship between inhibition of release and aggregation phenomena because aspirin depressed the ristocetin-induced release of ^{14}C-labeled serotonin in both LPC-treated PRP and in the control sample. One explanation for the effect of LPC is that it has modified the membrane environment around the cyclooxygenase, making it susceptible to aspirin inhibition. It is also possible that LPC causes a rearrangement of platelet receptors (glycoprotein Ib) inside the platelet membrane, rendering them more available to both the von Willebrand factor and the inhibitors.

Modification of Platelet Adhesive Properties by Lysolecithin

Another important platelet function to be considered is its interaction with blood vessel walls. As a model to study this process, we have examined the adhesiveness of platelets to glass, as described by Perlick.[13] A significant increase in the percentage of adhesive platelets was noted after incubation for 40 min in PRP containing LPC (350 μM) compared to saline controls (FIGURE 3). To examine how certain disease states influenced the effect of LPC on adhesion, platelets in PRP from patients suffering from cirrhosis of the liver and hyperlipemia-associated atherosclerosis were compared with those from control subjects (FIGURE 3). Without LPC, the

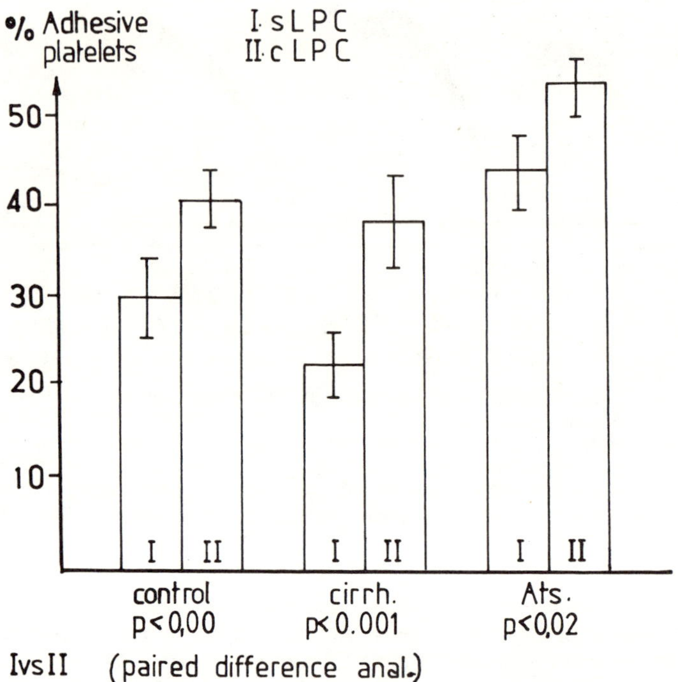

FIGURE 3. Effect of LPC upon platelet adhesiveness in control subjects, patients with liver cirrhosis (cirrh.), and atherosclerotic patients (Ats.). Results are the mean ± SEM.

platelets from cirrhotic individuals showed a lower adhesiveness than controls, whereas atherosclerotic individuals showed a higher adhesiveness. LPC exerted the greatest stimulation on the platelets from the cirrhotic individuals. One possible explanation could involve lecithin:cholesterol acyltransferase (LCAT) activity. This activity is reduced in severe liver disease[15] and hence the endogenous production of LPC is diminished. Platelets from cirrhotic patients may have membranes deficient in LPC or altered in some other way that makes them more susceptible to the effects of LPC.

Interactions Among Platelets, LPC, and Von Willebrand Factor

The adhesion of platelets to the subendothelial layers of the vessel wall in part depends on a plasma factor, the von Willebrand factor (VII R:WF), and involves a receptor on the platelet membrane, glycoprotein I (GPI). The involvement of these proteins is evident from deficiency diseases. Reduced adhesion of platelets to the subendothelium occurs in patients with a deficiency in factor VIII (von Willebrand's disease) and those whose platelets lack glycoprotein I (Bernard-Soulier syndrome).[16] A clear separation of aggregation and adhesion properties occurs in these diseases as platelets aggregate normally with ADP in spite of their impaired adhesive properties.

Evidence from our laboratory emphasizes that the plasma level of VIII R:WF was normal in hyperlipemic patients without clinical atherosclerosis, moderately ele-

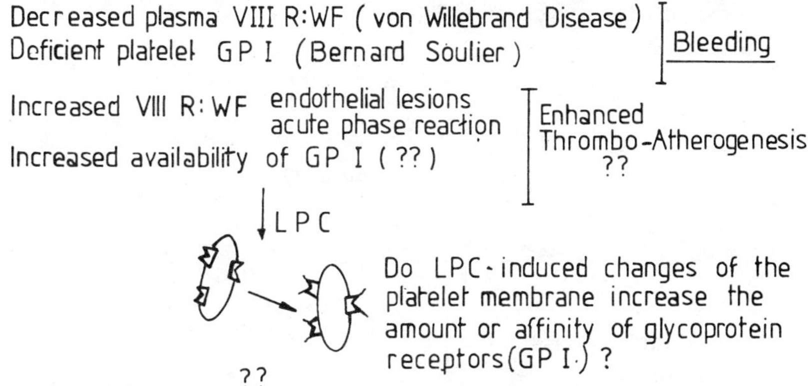

FIGURE 4. Schematic representation of possible disturbances of the interaction between platelet receptors (glycoprotein I, GP I) and endothelium-produced von Willebrand factor (VIIIR: WF).

vated in atherosclerotic patients with an obvious progression of the clinical symptoms, and greatly increased in patients with myocardial infarction, in postoperative conditions, and in toxemia of pregnancy.[17] These results suggest that factor VIII is involved in the progression of atherosclerosis. The possible relationships between platelet diseases, factor VIII, GPI, and LPC are illustrated in FIGURE 4. In view of the implications of an altered receptor function affecting platelet functions, the interaction between LPC and the glycoprotein I receptor warrants further attention. It is possible that high levels of factor VIII and an increased availability of the glycoprotein receptors might be particularly relevant for thromboatherogenesis.

The Relationship Between Hyperlipemia and Thrombotic Complications

Clinical evidence indicates that obesity, hypertension, diabetes mellitus, hyperuricemia, and vascular disease are frequently associated with a marked tendency towards thrombotic complications.[18-20] These disorders seem to occur mainly in hypertriglyceridemic patients,[20] but the detailed mechanisms leading to their association are still poorly understood. One of the main focuses in our laboratory has been to attempt to establish a relationship between hyperlipoproteinemia (HLP) and thrombotic tendency.

The liver is a vital organ in both lipoprotein synthesis and the synthesis of hemostatic factors. A working hypothesis relating relevant liver functions is illustrated in FIGURE 5. The levels of various liver-produced plasma factors, such as vitamin K-dependent clotting factors and fibrin-stabilizing factor XIII, have been shown to significantly increase in hyperlipidemic subjects, especially in those with endogenous hypertriglyceridemia.[21-23] It has also been established that the increased resistance to fibrinolysis of the clots from diluted blood obtained from hypertriglyceridemic subjects is at least partially caused by the high level of plasma factor XIII and by the subsequent enhanced binding of the α_2 plasmin inhibitor to the fibrin network.[24,25] Since other liver secretion enzymes (such as serum cholinesterase[22] and LCAT[26,27]) were also found to be increased in patients with hyperlipoproteinemia (HLP) types IIb and IV, the changes in hemostatic factors could be the result of enhanced hepatic

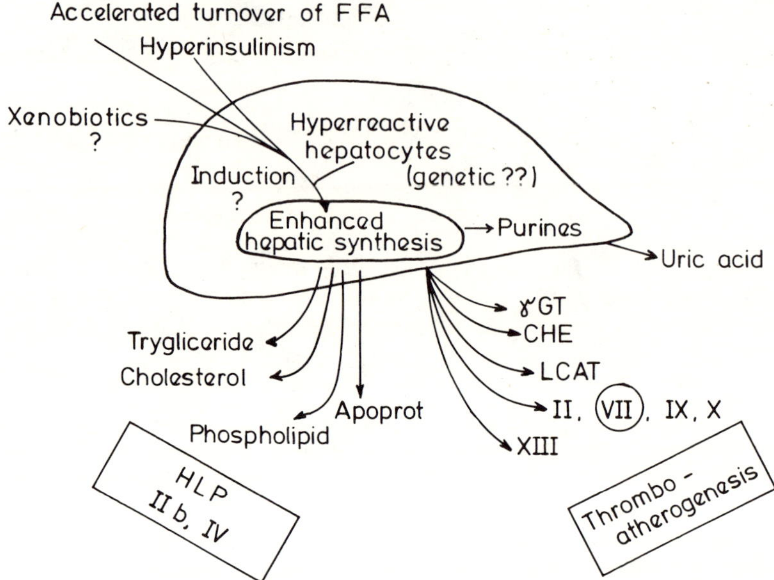

FIGURE 5. The possible relationship in liver between endogenous hypertriglyceridemia, accelerated turnover of the lipoproteins, induction of several hepatic enzymes and plasma proteins, and thrombotic tendency.

syntheses. Endogenous hypertriglyceridemia and increased synthesis of liver secretion enzymes could be different effects of a common cause. That is a response of genetically hyperreactive hepatocytes to various stimuli and inducing agents, which might be represented by xenobiotics as well as by endogenous metabolites and hormones.

Since one of the products of the LCAT-catalyzed reaction is LPC, an increased synthesis of LCAT may produce an increase in serum LPC. An increased concentration of LPC was reported in the aorta and serum of squirrel monkeys with nutritionally induced atherosclerosis. The atherosclerotic monkeys were also found to display a higher LCAT activity and a modified distribution of LPC within the various lipoprotein classes.[28,29] This supports a hypothesis that LPC might represent one of the pathogenic links between disturbances of lipid metabolism, abnormal platelet behavior, and thromboatherogenesis.

Data concerning the behavior of platelets in the various types of hyperlipoproteinemia (HLP) are rather conflicting. A shorter *in vivo* survival time of labeled platelets was reported in hypertriglyceridemic patients, but not in pure hypercholesterolemia.[30] In the latter patients an increased platelet aggregability was found.[31,32] As mentioned earlier, results obtained in our laboratory indicate that platelet adhesiveness to glass is enhanced mainly in those HLP patients with obvious vascular disease.[33] These results suggest that abnormal platelet behavior is related to lesions of the arterial wall rather than to a peculiar pattern of serum lipoproteins. A shortened *in vivo* platelet-survival time was noted in patients with arterial thrombosis[34] and an enhanced *in vitro* responsiveness to ADP was reported mainly in atherosclerotic patients with a high frequency of arterial thrombi, particularly in periods with obvious progression of the clinical symptoms.[34] It was also suggested that endothelial damage and disturbed blood flow might be accompanied by an enhanced removal of circulating platelets coupled with a subsequent accelerated production of young platelets with increased reactivity.[36] Any relationship between HLP and abnormal platelet behavior might therefore be indirectly determined by endothelial lesions resulting from a sustained atherogenic hyperlipoproteinemia.

This point of view is supported by experimental data obtained by cross transfusions of labeled platelets in hypercholesterolemic-atherosclerotic and normolipidemic monkeys. Such experiments demonstrated that the reduction in circulating platelet survival time is mainly caused by endothelial lesions subsequent to hypercholesterolemia and not by a direct effect of plasma lipids upon the circulating platelets.[37]

Rather few clinical investigations exist concerning the behavior of plasma LPC in cardiovascular disease associated with thromboatherogenesis. LCAT activity was found to be increased in endogenous hypertriglyceridemia,[26,27] and an enhanced production of LPC is to be expected in such a condition. Plasma LPC levels were elevated in the nephrotic syndrome, which is usually accompanied by HLP.[38] The acute-phase reaction occurring in myocardial infarction leads to a decrease in LCAT activity[39] and in plasma LPC levels.[40] Severe liver disease is also characterized by low LCAT activity and decreased LPC concentration in plasma.[15] Therefore, the acute-phase reaction and liver disease should be excluded when considering the low levels of plasma LPC reported in cardiovascular patients.[7,8]

In considering plasma LPC levels it must be remembered that LPC is produced not only through the LCAT-mediated reaction but also by phospholipase action on plasma lecithin.[41] Also, plasma LPC levels represent the balance between its production and its removal, by cell or tissue uptake or by its metabolism to PC or glycerophosphorylcholine.

The above evidence indicates that further studies of LPC turnover in HLP and cardiovascular disease are required. Such studies should be directed at the role of LPC in influencing various membrane functions in platelets.

Acknowledgments

I am grateful to Professor J. F. Mustard from the Department of Pathology, McMaster University, Hamilton, Ontario, Canada, who introduced me to the study of LPC and platelets in 1968 and offered me facilities for the study of the effect of LPC upon ristocetin-induced platelet aggregation in 1981.

References

1. Weltzien, H. U. 1979. Cytolytic and membrane-perturbing properties of lysophosphatidylcholine. Biochim. Biophys. Acta **559**: 259–287.
2. McKean, M. L., J. B. Smith & M. J. Silver. 1981. Regulation of fatty acid composition of platelet phospholipids (Abstract 997). *In* Thrombosis and Haemostasis. 46 (7): 321. Abstracts of the VIII International Congress on Thrombosis and Haemostasis (Toronto, July 1981).
3. Kunze, H. & W. Vogt. 1971. Significance of phospholipase A_2 for prostaglandin formation. Ann. N.Y. Acad. Sci. **180**: 123–125.
4. Joist, J. H., G. Dolezel, J. V. Lloyd & F. Mustard. 1976. Phospholipid transfer between plasma and platelets *in vitro*. Blood **48**: 199–211.

5. Bolton, C. H., J. R. Hampton & J. R. A. Mitchell. 1967. Nature of the transferable factor which causes abnormal platelet behaviour in vascular disease. Lancet **11**: 1101–1105.
6. Hampton, J. R. & C. H. Bolton. 1969. Effect of phospholipids on platelet electrophoretic mobility. J. Atheroscler. Res. **9**: 131–139.
7. Besterman, E. M. M. & M. P. T. Gillet. 1971. Inhibition of platelet aggregation by lysolecithin. Atherosclerosis **14**: 323-330.
8. Gillet, M. P. T. & E. M. M. Besterman. 1974. A possible thromboprotective role for lysolecithin in man. Thromb. Res. **4**(suppl. 1): 85–86.
9. Joist, J. H., G. Dolezel, M. P. Cucuianu, E. E. Nishizawa & J. F. Mustard. 1977. Inhibition and potentiation of platelet function by lysolecithin. Blood **49**: 101–112.
10. Bekemeir, H., H. A. Bekathir, A. J. Giessler, R. Hirschelmann & P. Nuhn. 1981. Effect of lysolecithin on blood vessel reactivity and of some ethers and esters of glycerophosphocholine and phosphocholine. Resp. on aggregation of rat platelets. Les Colloques de L'INSERM. *In* Pharmacology of Inflammation. INSERM **100**: 501-505.
11. Jenkins, C. S. P., D. Meyer, M. D. Dreyfus & M. J. Larrieu. 1974. Willebrand factor and Ristocetin. I. Mechanism of ristocetin-induced platelet aggregation. Br. J. Haematol. **28**: 561–578.
12. Allain, J. P., H. A. Cooper, R. H. Wagner & K. M. Brinkhous. 1975. Platelets fixed with paraformaldehyde: A new reagent for assay of von Willebrand factor and platelet aggregating factor. J. Lab. Clin. Med. **85**: 318-328.
13. Geratz, J. D., R. R. Tidwell, K. M. Brinkhous, S. F. Mohammad, O. Dann & H. Loewe. 1978. Specific inhibition of platelet agglutination and aggregation by aromatic amidino compounds. Thrombos. Haemostas. **39**: 411–425.
14. Perlick, E. 1960. Gerinnungslaboratorium in Klinik und Praxis. G. Thieme. Leipzig.
15. Gjone, E. & K. R. Norum. 1970. Plasma lecithin-cholesterol acyltransferase and erythrocyte lipids in liver disease. Acta Med. Scand. **187**: 153–161.
16. Weiss, H. J., V. T. Turitto & H. R. Baumgartner. 1978. Effect of shear rate on platelet interaction with subendothelium in citrated and native blood. I. Shear rate-dependent decrease of adhesion in von Willebrand's disease and the Bernard-Soulier syndrome. J. Lab. Clin. Med. **92**: 750–764.
17. Cucuianu, M. P., I. Missits, N. Olinic & S. Roman. 1980. Increased ristocetin-cofactor in acute myocardial infarction: a component of the acute phase reaction. Thrombos. Haemostas. **43**: 41–44.
18. Bjorck, G. 1965. On the definition of atherosclerosis in clinical studies. J. Atheroscler. Res. **5**: 261–275.
19. Moga, A., M. Cucuianu, I. Orha & R. Vlaicu. 1970. Hipertensiunea arterială si ateroscleroza. Ed Medicală. Bucuresti.
20. Haller, H., W. Leonhardt, M. Hanefeld, I. Schulze & Th. Fritz. 1973. Bedeutung und Therapie des Risikofaktors Hyperlipoproteinämie. Symposium über Lipidstoffwechselerkrankungen. (Dresden, 1973) pp. 264–276.
21. Constantino, M., C. Merskey, D. I. Kudzma & M. B. Zucker. 1977. Increased activity of vitamin K-dependent clotting factors in human hyperlipoproteinemia. Association with cholesterol and triglyceride levels. Thrombos. Haemostas. **38**: 465–477.
22. Cucuianu, M., V. A. Vasile, T. A. Popescu, A. Opincaru, I. Crîsnic & D. Tapalaga. 1973. Clinical studies concerning factor XIII, with special reference to hyperlipemia. Thrombos. Diathes. Haemorrh. **30**: 480–493.
23. Cucuianu, M., K. Miloszewski, D. Porutiu & M. S. Losowsky. 1976. Plasma factor XIII and platelet factor XIII. Thrombos. Haemostas. **36**: 542–550.
24. Cucuianu, M., C. Stef, D. Zdrenghea & O. Popescu. 1979. *In vitro* effect of p-chlormercuribenzoate upon dilute blood clot lysis time in hyperlipemia. Thrombos. Haemostas. **42**: 929–944.
25. Sakata, Y. & N. Aoki. 1980. Crosslinking of α_2 plasmin inhibitor to fibrin by fibrin stabilizing factor. J. Clin. Invest. **65**: 290–297.
26. Rose, H. G. & J. Juliano. 1976. Regulation of plasma lecithin:cholesterol acyltransferase in man. I. Increased activity in primary hypertriglyceridemia. J. Lab. Clin. Med. **88**: 29–43.
27. Cucuianu, M., A. Opincaru & D. Tapalagă. 1978. Similar behaviour of lecithin:

cholesterol acyltransferase and pseudocholinesterase in liver disease and hypercholes terolemia. Clin. Chim. Acta **85**: 73–79.
28. Portman, O. W. & M. Alexander. 1969. Lysophosphatidylcholine concentrations and metabolism in aortic intima plus inner media: effect of nutritionally-induced atherosclerosis. J. Lipid Res. **10**: 158–165.
29. Portman, O. W., P. Soltys, M. Alexander & T. Osuga. 1970. Metabolism of lysolecithin *in vivo*: effects of hyperlipemia and atherosclerosis in squirrel monkeys. J. Lipid Res. **11**: 596–604.
30. Jonker, J. J. C., M. R. Veen, W. Schopman & G. J. C. den Ottolander. 1974. Platelet survival time in angina pectoris and hyperlipoproteinemia. Thrombos. Res. **4**(suppl. 1): 65–67.
31. Carvalho, C. A., R. W. Colman & R. S. Lees. 1974. Platelet function in hyperlipoproteinemia. N. Engl. J. Med. **290**: 434–438.
32. Colman, R. W. 1978. Platelet function in hyperbetalipoproteinemia. Thrombos. Haemostas. **39**: 284–293.
33. Zdrenghea, D., C. Stef, M. Cucuianu & R. Vlaicu. 1979. Behaviour of platelet adhesiveness and of dilute blood clot lysis time in hyperlipoproteinemic subjects. Rev. Roum. Med. Int. **17**: 273–277.
34. Harker, L. A. & J. J. Schlichter. 1972. Platelet and fibrinogen consumption in man. N. Engl. J. Med. **287**: 999–1005.
35. Gormsen, J., J. Dalsgaard-Nielsen & L. Agerskov-Andersen. 1977. ADP-induced platelet aggregation *in vitro* in patients with ischaemic heart disease and peripheral thromboatherosclerosis. Acta Med. Scand. **201**: 509–513.
36. Mustard, J. F. & M. A. Packham. 1977. Platelets and diabetes mellitus. N. Engl. J. Med. **297**: 1345–1347.
37. Ross, R. & L. Harker. 1976. Hyperlipidemia and atherosclerosis; chronic hyperlipidemia initiates and maintains lesions by endothelial cell desquamation and lipid accumulation. Science **193**: 1094–1100.
38. Nye, W. H. C. & C. Waterhouse. 1961. The phosphatides of human plasma. II. Abnormalities encountered in the nephrotic syndrome. J. Clin. Invest. **40**: 1202–1207.
39. Ritland, S., J. P. Blomhoff, S. C. Enger, S. Skrede & E. Gjone. 1975. The esterification of cholesterol in plasma after acute myocardial infarcation. Scand. J. Clin. Lab. Invest. **35**: 181–187.
40. Berlin, R., C. O. Oldfeld & O. Vikrot. 1969. Acute myocardial infarction and plasma phospholipid level. Acta Med. Scand. **185**: 439–442.
41. Berlin, R., C. O. Oldfeld & O. Vikrot. 1969. Increased lecithinase activity after heparin administration. Acta Med. Scand. **185**: 433–437.

ERYTHROPOIETIN INTERACTION WITH THE MATURE RED CELL MEMBRANE

I. Baciu and Liana Ivanof

Department of Physiology
Institute of Medicine and Pharmacy

Physiology Laboratory
Institute of Hygiene and Public Health
Cluj-Napoca, Romania

Erythropoietin is a glycoprotein hormone synthesized primarily in the kidney in response to a decline in the oxygen content of arterial blood.[1-4] Large amounts are produced in response to hypoxia and other oxygen-deficiency states. The hormone promotes the proliferation and differentiation of medullary red cells, hemoglobin synthesis, and reticulocyte release from bone marrow.

Mature, anucleated erythrocytes were thought to be insensitive to erythropoietin. However, when we examined ribose uptake in mature cells, an activity that is increased in immature cells by erythropoietin, it too was observed to increase in response to erythropoietin in a concentration-dependent manner.[5] This finding prompted us to study in more detail the interaction of erythropoietin with red cell membranes.[6] We have studied the distribution of erythropoietin in blood between serum and the erythrocyte surface, the binding characteristics of erythropoietin to the erythrocyte membrane, and some properties of the membrane receptors for erythropoietin.

Experimental Procedures

Endogenous erythropoietin secretion was induced by hypobaric hypoxia (0.5 atm) of various durations, by hemolytic anemia produced with phenyl hydrazine, by repeated bleeding, or by the administration of exogenous erythropoietin. This erythropoietin was extracted either from the urine of patients with hypoplastic anemia or from the serum of sheep bled to a hematocrit of 12–15%, using methods previously described.[7] The isolated extracts contained 2–20 U/mg.

Erythropoietin activity was estimated by two methods. The increase in the rate of ^{59}Fe incorporation into red cells following the injection of extracts into rats was used as described by Plzak *et al.*[8] The other method, developed in our laboratory, relies on the stimulation of ribose uptake by red cells *in vitro* in response to the hormone (I. Baciu and Liana Ivanof, unpublished). The effect of erythropoietin on ribose uptake is illustrated in Figure 1. The results of these two tests are essentially identical (Figure 2) where erythropoietin release in rats in response to hypoxia was followed. In some experiments, both assays were used, whereas in others only one was used.

Results

The Distribution of Erythropoietin Between Serum and Red Cells

Binding of erythropoietin to erythrocytes was determined *in vivo* in hypoxic rats (exposed to 0.5 atm for 22 hours). When the hematocrit level was raised by hyper-

0077-8923/83/0414-0066$01.75/0 © 1983, NYAS

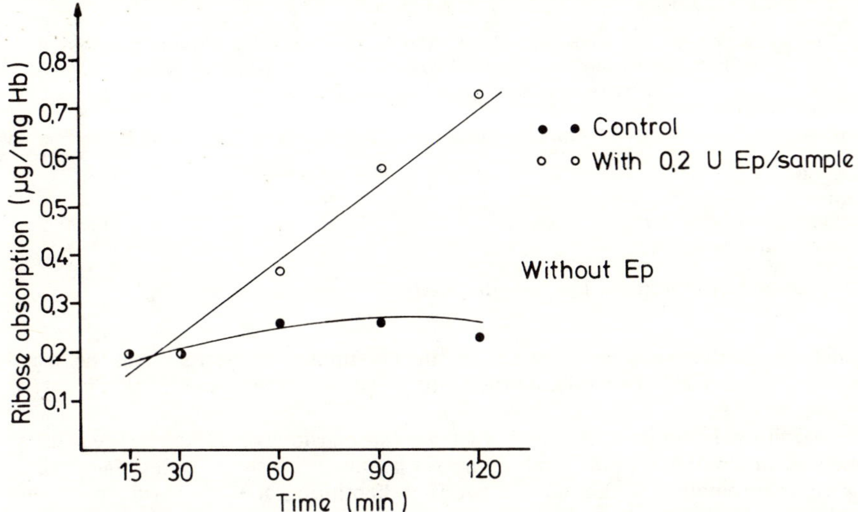

FIGURE 1. Effect of erythropoietin (Ep) on ribose uptake by erythrocytes.

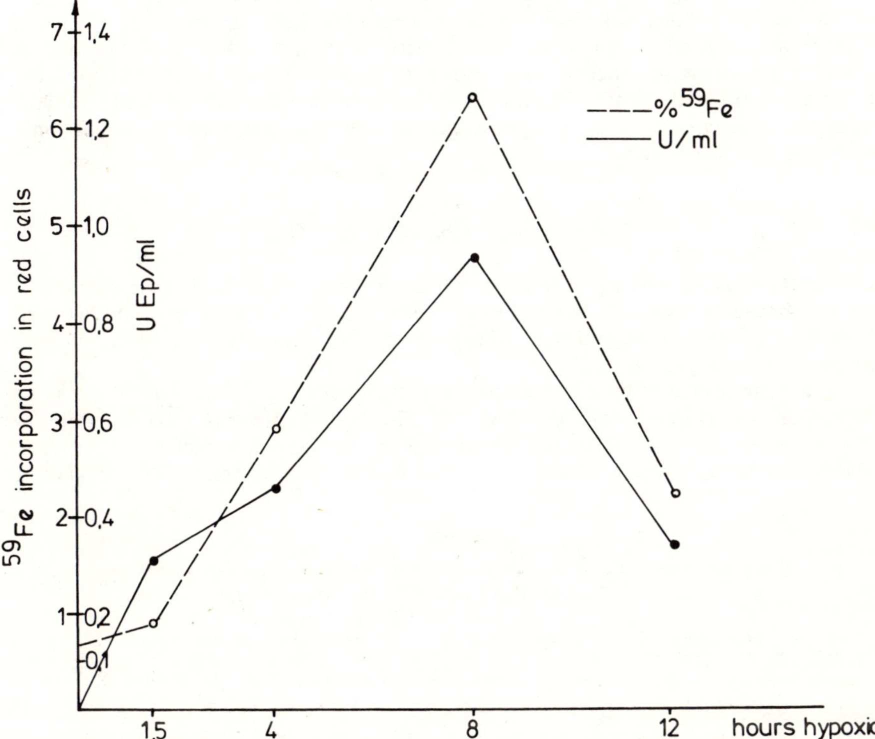

FIGURE 2. A comparison of ^{59}Fe incorporation and ribose uptake by erythrocytes in hypoxic rat serum, where ribose uptake has been converted to U Ep/ml by reference to a standard curve.

TABLE 1

THE RELATIONSHIP BETWEEN THE HEMATOCRIT LEVEL AND THE CONCENTRATION OF SERUM ERYTHROPOIETIN IN HYPOXIC RATS AND HYPOXIC RATS HYPERTRANSFUSED WITH NORMAL ERYTHROCYTES

Controls	No. of Animals	Hematocrit %	U Ep/ml Serum
Controls	7	45.5	ND
Hypoxic	6	48.8	0.85
Hypoxic hypertransfused	6	60	0.675
Hypoxic hypertransfused	8	68	ND

N.D. = Not Detectable, Ep = erythropoietin.

transfusion with normal erythrocytes, erythropoietin activity decreased as the hematocrit level increased (TABLE 1). At the highest hematocrit level used, 68%, no activity was detected.

Additional data were obtained *in vitro* on suspensions of human red cells in hypoxic serum containing 3.6 U/ml at various hematocrit levels. Testing the erythropoietin that remained in the supernatant after 30 minutes at 37°C, it was found that the hormone concentration decreased as the hematocrit increased (TABLE 2).

This distribution of erythropoietin between the incubation medium and erythrocyte suspensions *in vivo* and *in vitro* is in agreement with the findings of other research workers. Thus, Adamson,[9] Adamson and Finch,[10] Hammond *et al.*,[11] and Alexanian[12] reported a negative correlation between hematocrit levels and plasma or urinary erythropoietin concentrations; however, such a correlation was not apparent in polycythemia and anemia. We consider that these data can be explained by erythropoietin binding to red cells. A significant rise in plasma erythropoietin occurs only at low hematrocrit levels.

Erythropoietin Binding to Red Cells

Erythropoietin binding to red cells could represent a nonspecific attachment to the membrane or a specific binding to a receptor. In order to discriminate between these two possibilities, several types of experiments were carried out *in vitro* on intact erythrocytes and ghosts.

The time course of binding was followed by incubating erythrocyte suspensions (hematocrit 41%) in hypoxic serum at 37°C. Erythropoietin was determined in the erythrocyte supernatant at 30 and 60 minutes. From the initial value of 3.9 U/ml, erythropoietin decreased to 2.2 U/ml at 30 min and disappeared totally by 60 min.

TABLE 2

ERYTHROPOIETIN CONCENTRATION IN SERUM FOLLOWING INCUBATION OF SUSPENSIONS OF HUMAN ERYTHROCYTES OF VARIOUS HEMATOCRITS

Hematocrit %	U Erythropoietin/ml Serum
–	3.6
33	2.48
42	2.48
55	1.74
74	0.30

The saturation of binding sites for erythropoietin on erythrocytes was shown by incubating hypoxic red cells in hypoxic serum. In 30 min at 37°C, normal erythrocytes were able to reduce the hormone concentration in hypoxic serum from 1.1 U/ml to 0.2 U/ml. Erythrocytes isolated from hypoxic rats (5 hours at 0.5 atm) however, could reduce the hormone concentration to only 0.8 U/ml. This indicates that normal erythrocytes have a greater capacity to bind erythropoietin than erythrocytes isolated from hypoxic blood, and suggests that the bulk of the erythropoietin-binding sites are occupied on hypoxic erythrocytes because of their exposure to the hormone released during the hypoxia.

The limited binding of erythropoietin to the hypoxic erythrocytes also suggests that the hormone is not rapidly metabolized by erythrocytes, if at all. That the bound erythropoietin remains active was shown by analyzing ribose uptake by control cells after the 30-min incubation with hypoxic serum. The initial rate of ribose uptake increased from 0.277 μg/mg hemoglobin/hr to 0.455 μg/mg hemoglobin/hr.

The effect of temperature on binding is illustrated in FIGURE 3. Substantial binding occurred at 0°C and binding increased with increasing temperature. As binding of ligands to receptors at low temperature is normally indicative of non-specific binding, the effect of temperature suggests that the binding may consist of specific and non-specific components.

Bound erythropoietin could be released from erythrocytes by lowering the pH to 6.5. Ninety percent was recovered following incubation at 4°C in 150 mM NaCl, pH 6.5, for 12 hours. This indicates that the eluted erythropoietin preserves the bulk of its activity. Transfer of bound erythropoietin and the preservation of its activity were demonstrated *in vivo*. Hormone was bound to red cell membranes (resealed ghosts) by incubation at 37°C for 30 min. Unbound erythropoietin was removed from the ghosts by three washings with 5 mM phosphate buffer (pH 8.0). The mem-

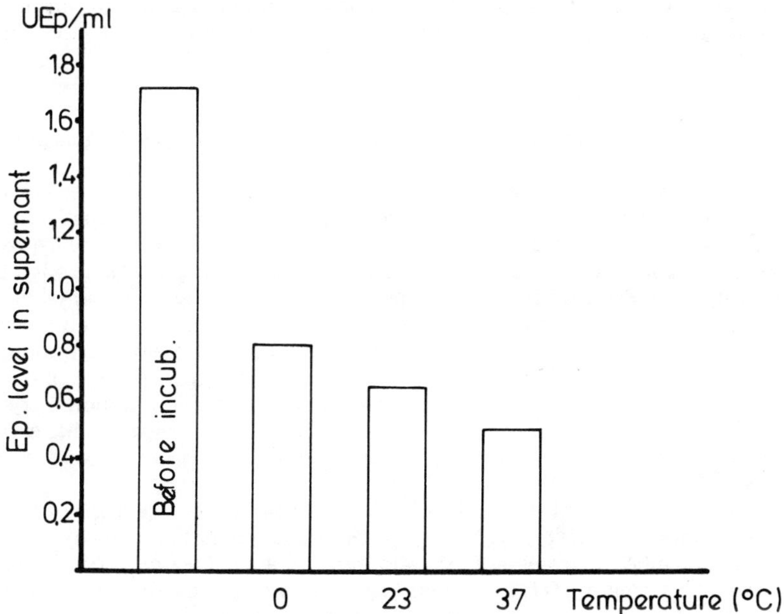

FIGURE 3. The effect of incubation temperature on the binding of Ep to erythrocytes.

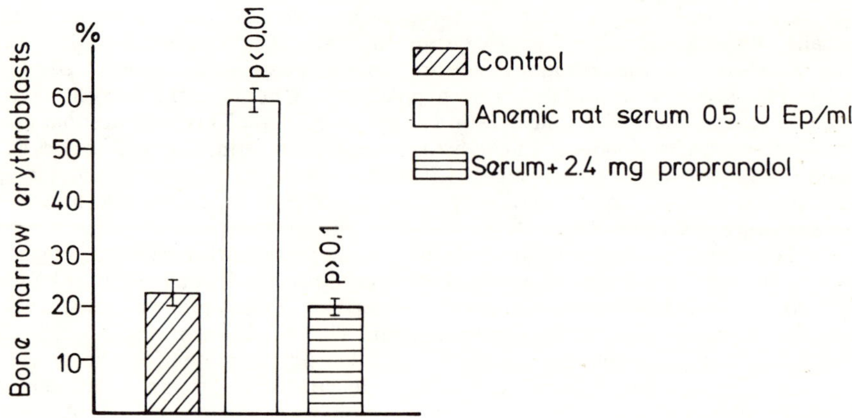

FIGURE 4. Stimulation of erythroblast formation and its inhibition by propranolol following injection of 0.5 U Ep daily for three consecutive days. The statistical difference to controls was calculated.

branes were resuspended in phosphate-buffered saline at a concentration of approximately 1 mg protein/ml. Amounts of 0.5 mg membrane protein were injected intravenously in mice. After three hours, blood was taken from the orbital sinus and tested for ribose uptake by the red cells. An increase in ribose incorporation of 41% was found compared to the control animals injected with nonadsorbed membranes. This demonstrates transfer of bound erythropoietin from resealed ghost membranes to erythrocytes *in vivo* and the preservation of erythropoietin activity after binding to the ghosts.

Preliminary Characterization of Mature Red Cell Receptors for Erythropoietin

Preliminary reports indicated that salbutamol, a beta-2 antagonist, had erythropoietic effects, while the beta-blocking agents (propranolol) or beta-specific antagonists (butoxamine) lowered the production of erythropoietin induced by hypoxia.[13] These experiments suggest that erythropoietin acts on the target cells through beta-adrenergic-type membrane receptors. Similar results were obtained in our laboratory with mice when erythropoietin (0.5 U daily) and propranolol (2.4 mg daily) were administered simultaneously on three consecutive days (I. Baciu, Cornelia Marina, and Coleta Zdrenghea, unpublished results). As shown in FIGURE 4, propranolol inhibited erythropoiesis. Propranolol also blocked ^{59}Fe incorporation in mouse erythrocytes following injection of anemic-rat serum (FIGURE 5). This suggests that erythropoietin acts through a beta-adrenergic receptor in mature cells similar to the type of receptor active during erythropoiesis.

The protein nature of the binding sites was confirmed by treatment of erythrocyte membranes[14] with trypsin (100 μg/mg membrane protein) or with chymotrypsin (200 μg/mg membrane protein), which abolished binding. Treatment of membranes with low ionic strength EDTA solutions (0.1 mM), which removes peripheral membrane proteins,[14] did not affect binding. This indicates that the receptor is an integral membrane protein, probably spanning the membrane.

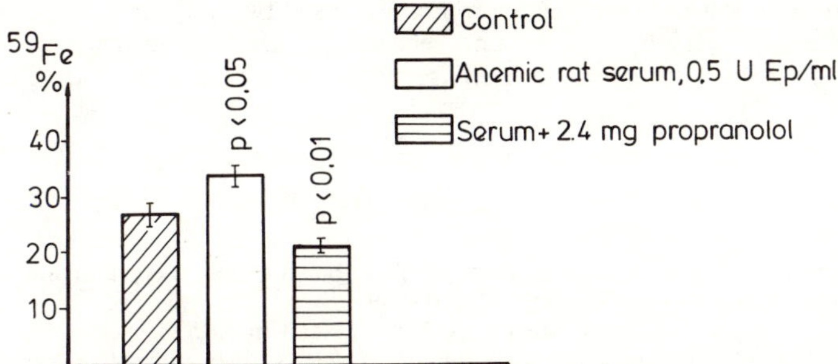

FIGURE 5. Stimulation of ^{59}Fe incorporation and its inhibition by propranolol in mouse erythrocytes isolated from mice injected with anemic rat serum (0.5 U Ep/ml). The statistical difference to the controls was estimated.

Discussion

Erythropoietin is designated as the hormone stimulating erythroid cell mitoses. The medullary nucleated elements, forming a continual population, starting with the stem cells and continuing with erythroblasts, respond to erythropoietin by the synthesis of nucleic acids. The normoblasts, reticulocytes, and erythrocytes that do not divide any more do not exhibit this response.

Goldwasser[4] proposed that erythropoietin may initially interact with receptors on erythroid cells and that the signal may be transmitted by means of a cytoplasmic factor. This factor, presumably a protein, may be formed or released under the action of the hormone and would initiate a specific nuclear transcription. Experimental evidence supports this type of hormone-receptor interaction with stem cell membranes and the membranes of the erythroid cells.[15,16] Our data with beta-blockers indicate that the same type of interaction occurs between erythropoietin and erythroblast membranes, although the response is different.

The effect of erythropoietin on the erythrocyte membrane is unusual in that red cells lack the capacity to synthesize intracellular proteins and to replicate. Erythropoietin binding to mature cells could have at least two functional consequences. The first is that binding to the cell may act to remove hormone excess from the plasma, preventing its desialization and inactivation by tissues. As bound erythropoietin is not available to act on bone marrow cells, it may constitute a regulatory mechanism. Furthermore, as binding is reversible, it may be slowly released to stimulate erythropoiesis.

The second consequence derives from the role of ribose as a precursor in the synthesis of nucleotides and 2,3-diphosphoglycerate (DPG) in erythrocytes. In our laboratory we are presently studying whether this hormone modulates DPG synthesis in mature red cells by increasing ribose uptake and utilization. DPG is known to be important in modifying hemoglobin affinity for oxygen. In instances of tissue oxygen deprivation including cardiac failure, anemia, and arterial oxygen desaturation, red cells have been observed to respond with an increase in intracellular DPG.[1] This response has been postulated to involve DPG binding to hemoglobin, a decreased intracellular pH, and an increased rate of glycolysis.[1] An attractive extension of this proposal is that erythropoietin synthesis, in response to oxygen deprivation, results in enhanced binding of erythropoietin to erythrocytes and a stimulation of intracellular metabolism. Thus erythropoietin reacts in two ways to a decreased arterial oxygen: firstly, in erythropoiesis, and secondly, in decreasing hemoglobin affinity for oxygen. Both responses will increase tissue oxygenation but on different time scales.

References

1. Adamson, J. W. & C. A. Finch. 1975. Hemoglobin function, oxygen affinity, and erythropoietin. Annu. Rev. Physiol. **37**: 351–369.
2. Fisher, J. W. 1977. Kidney Hormones. Vol. 2. Academic Press. New York.
3. Baciu, I. 1980. Homeostazia oxigenului. Ed. Dacia. Cluj-Napoca, Romania.
4. Goldwasser, E. 1981. Erythropoietin and red cell differentiation. Control of Cellular Division and Development. Part A: 487–494.
5. Baciu, I., L. Ivanof, A. Muresan & T. Pavel. 1977. Influenta eritropoetinei asupra absorbtiei de riboza de catre celulele maduvei osoase. Clujul Medical **50**: 266–272.
6. Baciu, I., L. Ivanof & T. Pavel. 1980. Effect of erythropoietin on ribose consumption in rat erythrocytes. Clujul Medical **53**: 47–54.
7. Baciu, I., E. Rosenfeld, M. Dorofteiu & M. Rusu. 1959. Stimularea eritropoezei prin hipoxie. Nota III. Izolarea si proprietatile biologice ale unui factor proteic de stimulare a eritropoezei. Studii si cercet. med. (Cluj) **10**: 47–62.
8. Plzak, L. F., W. Fried, L. O. Jacobson & W. F. Bethard. 1955. Determination of stimulation of erythropoiesis by plasma from anemic rats using ^{59}Fe. J. Lab. Clin. Med. **46**: 671–678.
9. Adamson, J. W. 1968. The erythropoietin/hematocrit relationship in normal and polycythemic man. Implications of marrow regulation. Blood **32**: 597–609.
10. Adamson, J. W. & C. A. Finch. 1968. Erythropoietin and the polycythemias. Ann. N.Y. Acad. Sci. **149**: 560–563.
11. Hammond, G. D., A. I. Ishikoma & G. Keighley. 1966. Relationship between erythropoietin and severity of anemia in hypoplastic and hemolytic states. *In* Erythropoiesis. L. O. Jacobsen & M. Doyle, Eds.: 351. Grune Stratton. New York.
12. Alexanian, B. 1977. Increased erythropoietin production in man. *In* Kidney Hormones. J. W. Fisher, Ed. **2**: 531–569. Academic Press. New York.
13. Beckman, B. & J. W. Fisher. 1979. Changes in beta-2 adrenergic receptor sensitivity with maturation of erythroid progenitor cells. Experientia **35**: 1671–1672.
14. Fairbanks, G., T. L. Steck & D. F. H. Wallach. 1971. Electrophoretic analysis of the major polypeptides of the human erythrocyte membrane. Biochemistry **10**: 2606–2617.
15. Chang, S. C. S. & E. Goldwasser. 1973. On the mechanism of erythropoietin induced differentiation. XII. A cytoplasmic protein mediating induced nuclear RNA synthesis. Dev. Biol. **34**: 246–254.
16. Chang, S. C. S., D. Sikkema & E. Goldwasser. 1974. Evidence for an erythropoietin receptor protein on rat bone marrow cells. Biochem. Biophys. Res. Commun. **57**: 399–405.

CONTROL OF HEART OXIDATIVE PHOSPHORYLATION BY CREATINE KINASE IN MITOCHONDRIAL MEMBRANES*

William E. Jacobus,† Randall W. Moreadith, and Koenraad M. Vandegaer

Peter Belfer Laboratory for Myocardial Research
Department of Medicine (Cardiology)
Department of Physiological Chemistry
The Johns Hopkins University School of Medicine
Baltimore, Maryland 21205

INTRODUCTION

In the beginning God created lipids and proteins. It really does not matter which came first. Spontaneously they joined forces to form membranes. Then God said: "Now it is time to labor and work." Some were set about to establish ion gradients, transport nutrients, and protect the cell from hostile chemical environmental attack. These became cell membranes. Others received permits to develop and evolve inside these newly established factories of life. They also learned how to pump ions (sarcoplasmic reticulum), contain the genetic material (nuclear membrane), trap degradative enzymes (lysosomal membrane), form the site for protein synthesis (endoplasmic reticulum), and perform coupled electron transport (mitochondria). The subject of this report is not the architecture or biophysics of membranes, but the biochemical role a membrane plays in the performance of an essential cell function, the control of oxidative phosphorylation at the inner mitochondrial membrane of eukaryotic cells. The question asked is straightforward. In a cell, how are the rates of energy production regulated so that the bioenergetic demands of the tissue are met?

The heart is an ideal organ for the study of this type of question. This is because more than 90% of the ATP required for cellular function is derived from the aerobic reactions of the mitochondrial electron transport chain.[1] Since there is only a minor contribution toward ATP production by glycolysis, measurements of the rates of tissue oxygen consumption quite closely reflect the rates of energy supply in the myocardium.[2] Over the years physiologists have defined the determinants of myocardial oxygen consumption (TABLE 1). These have conveniently been divided into major and minor determinants. The major factors include myocardial contractility, intramyocardial wall tension (blood pressure and ventricular volume), and heart rate. Minor determinants are fiber shortening (Fenn effect), activation energy, and basal (resting) metabolism. With respect to ATP utilization, it has been documented that these minor determinants constitute less than 10% of the ATP demand in the heart. Thus, the major energy demands of the heart can be quantitated by direct measurements of myocardial contraction. Therefore, the reason the heart is such an ideal organ for studying questions concerning the coordination of energy supply to cellular energy use is that both parameters can easily be measured, which makes integrated biochemical and physiological investigations possible.

* Supported by United States Public Health Service Grants HL-20658 and P50 HL-17655 for the Specialized Center for Ischemic Heart Disease.

† Established Investigator of the American Heart Association and author to whom correspondence should be addressed: Peter Belfer Laboratory, 538 Carnegie Bldg., Division of Cardiology, The Johns Hopkins Hospital, 600 N. Wolfe St., Baltimore, MD 21205.

0077-8923/83/0414-0073$01.75/0 © 1983, NYAS

TABLE 1

DETERMINANTS OF MYOCARDIAL OXYGEN CONSUMPTION

Major	Minor
Myocardial contractility	Fiber shortening
Intramyocardial wall tension	Activation energy
Heart rate	Basal metabolism

From the above discussion it is apparent that the rates of mitochondrial oxidative phosphorylation must be rigorously controlled to meet the ATP demands of the sarcoplasm. This phenomenon is called respiratory control, and how it is achieved is one of the critical issues in the field of cellular bioenergetics. For most tissues this problem probably involves the direct fluxes of the adenine nucleotides and inorganic phosphate. However, in tissues that contain creatine kinase (heart, brain, and skeletal muscle), a more complicated picture has emerged in the past few years. Work from numerous laboratories[3-7] has now shown that creatine kinase plays an important, if not essential, role in the control and distribution of high-energy phosphate metabolites. A model of this is shown in FIGURE 1. According to this diagram, the ATP produced by oxidative phosphorylation is transphosphorylated to phosphocreatine by the mitochondrial isozyme of creatine kinase.[8-12] Phosphocreatine then diffuses into the cytoplasm interacting with other isozymes of creatine kinase to regenerate ATP.[13-15] Where this occurs depends upon the cytoplasmic localization of the enzyme. In heart, more than 90% of the cytoplasmic creatine kinase is tightly associated with subcellular components. These include the sarcolemmal membrane,[16] the sarcoplasmic reticulum,[17-19] and the cardiac myofibrils.[20-22] At these sites, ATP may be locally regenerated to energize the critical reactions of the excitation-contraction-relaxation cycle. The focus of this presentation is upon the reactions occurring at the

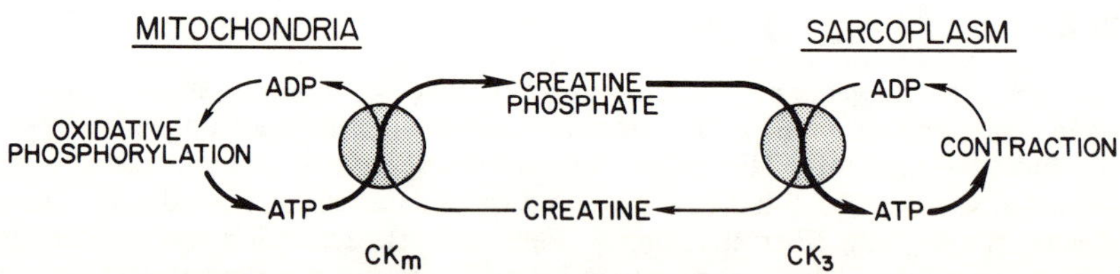

FIGURE 1. Model for heart energy transport. Matrix ADP is phosphorylated to ATP by the reactions of oxidative phosphorylation. The nucleotides are translocated across the inner mitochondrial membrane by the adenine nucleotide translocase. At the exterior surface of the inner membrane, the mitochondrial isozyme of creatine kinase (CK_m) interacts with ATP and creatine to form ADP and creatine phosphate. The ADP recycles into the matrix for rephosphorylation to ATP. Creatine phosphate diffuses in the sarcoplasm to interact with the bound cytoplasmic forms of creatine kinase (CK_3). At these sites, ADP is locally transphosphorylated to ATP, which is then utilized for the bioenergetic reactions of contraction, including ion transport and fiber shortening. To complete the cycle, creatine diffuses back to CK_m for rephosphorylation to creatine phosphate. In this model, creatine phosphate is a carrier or transport molecule, directly linking ATP demand to its production, a view quite different from that of it being a simple energy reservoir, utilized only in emergency conditions. (From Jacobus & Ingwall.[6] With permission from Williams & Wilkins.)

left side of the model (FIGURE 1), the control and integration of heart oxidative phosphorylation.

In heart mitochondria this appears to involve the mutual activities of an intrinsic membrane transport system, the adenine nucleotide translocase, and a peripheral membrane protein, the mitochondrial isozyme of creatine kinase. Several lines of investigation have suggested that these two proteins act as a multi-step enzyme complex, functionally coupled for the aerobic production of phosphocreatine.[23-25] This is a novel situation, for it represents the rare example of an enzyme complex involved in the dynamic coordination of a membrane transport protein and an enzyme. Even more interesting, however, is the fact that the adenine nucleotide translocase functions not only to supply substrate, ATP,[25-27] but also to remove product, ADP,[28,29] for the creatine kinase reaction. The data presented in this report suggest that the synergistic activity of these two proteins directly controls the rate of heart mitochondrial oxidative phosphorylation and, thus, myocardial oxygen consumption.

EXPERIMENTAL PROCEDURES

Materials

The adenine and pyridine nucleotides, 4-(2-hydroxyethyl)-1-piperazineethanesulfonic acid, ethylene glycol bis (*B*-aminoethyl ether)-*N*-*N*-*N*-*N*-tetraacetic acid, phosphoenolpyruvate, immobilized hexokinase, and the coupling enzymes were purchased from Sigma Chemical Co. (St. Louis, Mo.). The adenine nucleotides were the substantially vanadium-free grades. The enzymes were the following types: hexokinase, sulfate-free Type F-300 from yeast; lactic dehydrogenase, Type XI from rabbit muscle; glucose-6-phosphate dehydrogenase, sulfate-free Type IX from yeast; pyruvate kinase, Type III from rabbit muscle; and agarose-immobilized yeast hexokinase. The insoluble hexokinase was purchased in 25-IU vials containing 3 M ammonium sulfate. Prior to use the enzyme was processed as previously described.[28] Sucrose and glucose were obtained from J. T. Baker (Phillipsburg, N.J.), and all other reagents were of the highest purity commercially available. All solutions were prepared in deionized water.

Mitochondrial Isolations

Retired Sprague-Dawley breeder rats were used as the tissue source. The mitochondrial fractions from heart and liver were isolated according to a modification of the method of Vercesi *et al.*[30] The isolation medium contained 0.5 mM K-EGTA, 3.0 mM K-HEPES (pH 7.4), 0.25 M sucrose, and Sigma protease (1 mg/g, trimmed wet weight of heart) replaced Nagarse for the heart isolations; no protease was used in the liver mitochondrial isolations. Standard methods of differential centrifugation were employed,[31] and care was taken to maintain solutions at 2°C. The yield of mitochondrial protein averaged 10–12 mg per g heart. Respiratory control ratios varied from 4–9 for heart and 6–10 for liver (5 mM succinate and 5 μM rotenone). The ADP/O ratios were 1.4–2.2 for heart and 1.4–1.9 for liver fractions. The final mitochondrial pellets were resuspended at 50 mg protein/ml for heart and 25 mg protein/ml for liver. In both cases the fractions were stored at 2 to 4°C, and used within 3–4 hr of isolation. Protein concentrations were determined by the method of Bradford[32] with nitrogen-calibrated, crystallized bovine serum albumin used as the primary protein standard.

Oxygraph Procedures

Respiratory traces were obtained using a Clark oxygen electrode (Yellow Springs Instrument Co.) in a 3.0-ml water-jacketed oxygraph chamber maintained at 37°C by a Haake E12 circulation heater. Voltage changes were recorded with a Heath-Schlumberger Model SR-2558 recorder. The oxygraph medium contained 0.25 M sucrose, 3.0 mM HEPES, 2.0 mM KHP_i, 5.0 mM K-succinate, 0.5 mM EGTA, 11 mM $MgCl_2$, and 5 μM rotenone at pH 7.2. Stock solutions were maintained at 37°C. A value of 390 ng atoms of oxygen/ml was used for the solubility of oxygen at this temperature. Additional details concerning the sequence of additions are indicated in the figure legends. The protein concentrations were 0.15 mg/ml for heart and 0.5 mg/ml for liver.

Sample Extractions

Within 10 sec after the addition of hexokinase, steady-state rates of respiration were achieved and remained linear for an additional 2.0 min. At 1.0–1.5 min, a 2.5-ml sample was removed and added to 0.75 ml of 18% $HClO_4$. The mixture was immediately filtered using 0.45-μm Millipore filters and neutralized with 0.25 ml of 5 M K_2CO_3. The resultant crystals of potassium perchlorate were removed by centrifugation at 3000 × *g* for 15 min. The supernatant solution was decanted and frozen for nucleotide and P_i analysis.

Determination of ATP, ADP, and P_i

ATP was determined by a spectrophotometric assay previously described.[33] The content of ADP was assayed using a lactic dehydrogenase–pyruvate kinase spectrophotometric assay.[34] In both cases, the samples were corrected for matrix nucleotide content.[28]

Inorganic phosphate was measured by a colorimetric method.[35] A 0.1-ml sample of the extract was added to the molybdate mix and the color was allowed to develop for 10 min. The optical density of the solution was measured at 660 nm using a Gilford 2400 spectrophotometer.

Estimation of Apparent K_m *ADP*

To determine the apparent K_m values for ADP transport reported in TABLE 3, the measured ADP concentrations were plotted versus respiratory rates in double-reciprocal form, and the lines fit by the linear-regression program of the Hewlett-Packard 9810A calculator/plotter. In all cases, the R^2 values were greater than 0.90.

RESULTS AND DISCUSSION

Respiratory Control by ADP Availability

When intact mitochondria are suspended in oxygraph medium containing respiratory substrate, oxygen is consumed at a slow rate. This endogenous rate has been called State 4. It is well appreciated that in medium containing sufficient P_i, the addition of ADP leads to a marked acceleration in the rate of respiration, termed State 3. Upon completion of the phosphorylation of the added ADP to ATP, the rate of mitochondrial respiration decreases. If the mitochondria are well coupled, oxygen consumption returns to the initial State 4 rate. This simple phenomenon is called respiratory control. From the standpoint of cell physiology, most tissues consume oxygen at rates calculated to be faster than State 4, but significantly slower than State 3. The rates of tissue oxygen consumption can vary widely depending upon the physiological status of the tissue.[2] How these intermediate rates of respiration are achieved has been the subject of considerable debate for nearly 30 years.[36–59] Chance and Williams[36,37] in the mid 1950s proposed that the rates of respiration graded between State 3 and State 4 were a direct function of ADP availability. They reported that the apparent K_m for ADP-stimulated respiration was between 20–30 μM. Complications arose when people assayed the ADP content of tissue and found it to be much higher than the expected control range. Therefore, Klingenberg in 1961[38] postulated that the extramitochondrial phosphorylation potential, [ATP]/[ADP] × [P_i], was the metabolic parameter controlling the immediate rates of tissue oxygen utilization.

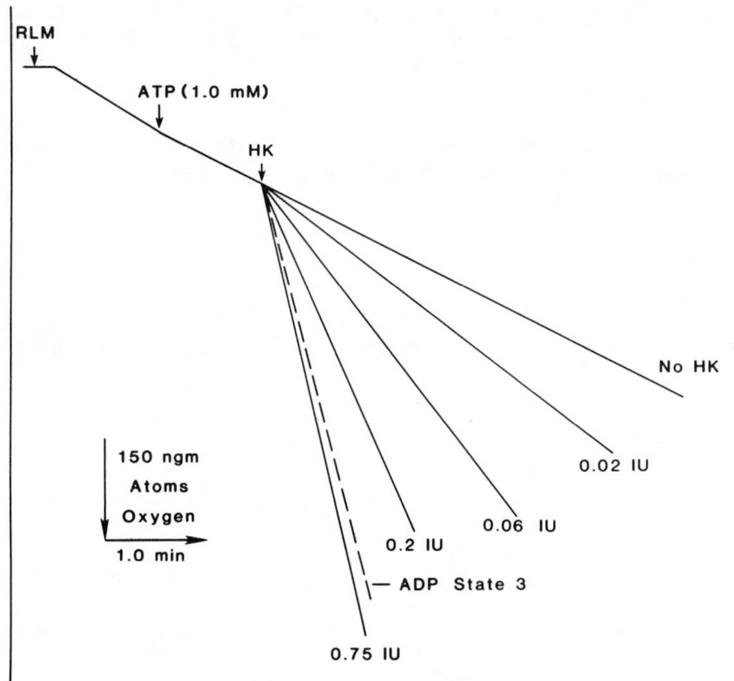

FIGURE 2. Hexokinase-induced graded respiratory rates. The 3.0-ml oxygraph chamber contained 0.25 M sucrose, 3 mM HEPES, 2 mM KHP_i, 5 mM K^+-succinate, 11 mM $MgCl_2$, 20 mM glucose, 5 μM rotenone at pH 7.2, 37°C. At the indicated points, 1.5 mg of rat liver mitochondrial protein (RLM), ATP (1.0 mM), and the indicated concentrations of yeast hexokinase (HK) were added. The State 4 rate of respiration is indicated by the No HK line. The State 3 rate, shown as the dashed line, was determined from the pulsed addition of 600 nmol of ADP into medium containing 1 mM ATP, but lacking glucose and hexokinase. (From Jacobus *et al.*[33] With permission from *Journal of Biological Chemistry*.)

This hypothesis has received much experimental support in the meantime.[39-49] Twelve years later, Slater *et al.*[50] postulated that respiratory control was a function of the [ATP]/[ADP] ratio, and thus somewhat independent of the concentration of P_i. This was based in part on the observation that ATP and ADP interacted at the adenine nucleotide translocase in a competitive manner. Therefore, kinetic control at the level of the translocase could modulate respiratory rates. This alternative theory has also been supported by a number of reports.[51-59] One of the classic approaches to the study of respiratory control is illustrated in FIGURE 2. This figure shows the oxygen-electrode traces when intact rat liver mitochondria (RLM) are incubated in oxygraph medium containing 20 mM glucose. At the indicated point, 1.0 mM ATP is added to provide substrate for the hexokinase reaction: ATP + Glucose → ADP + Glucose 6-Phosphate. To grade the steady-state rates of respiration, varying amounts of enzyme are added to the oxygraph chamber to initiate the hexokinase reaction and thus generate different steady-state concentrations of ADP. In this case, the rate of ADP production is limited by the quantity of added hexokinase. The ADP product is translocated into the mitochondrial matrix to stimulate oxidative phosphorylation. Therefore, as shown in FIGURE 2, by increasing the content of hexokinase, one observes faster rates of oxygen consumption. The addition of 0.75 IU to the 3.0-ml system leads to a rate that is just slightly faster than State 3. The data of FIGURE 3 show the complete titration curve for the rates of respiration versus enzyme content. We see that there is little or no further acceleration in respiration with enzyme additions above 0.75 IU. The respiratory capacity is saturated at this activity of hexokinase.

Using this protocol, we initiate the steady-state rates of respiration by enzyme addition. After waiting approximately 1.0 to 1.5 min, a 2.5-ml sample is removed, perchloric acid extracted, neutralized with potassium carbonate, centrifuged, and

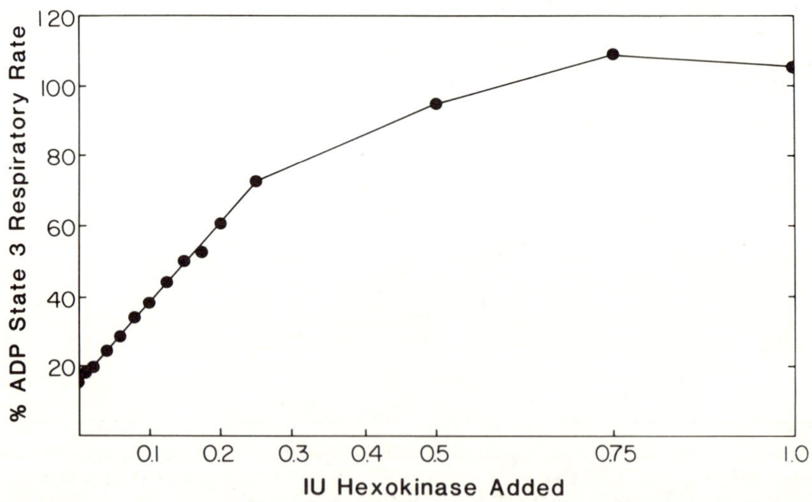

FIGURE 3. The complete titration curve for hexokinase-stimulated respiration. Conditions are as described in FIGURE 2. The abscissa represents the total IU of hexokinase added to the 3.0-ml medium. The ordinate is respiratory rate calculated as a percent of the control ADP-induced State 3 rate, measured as described in FIGURE 2. (From Jacobus *et al.*[33] With permission from *Journal of Biological Chemistry*.)

then supernatant assayed for the content of ATP, ADP, and P_i, by methods described in EXPERIMENTAL PROCEDURES. From these analytical data, one then calculates the phosphorylation potential values or the [ATP]/[ADP] ratios. These are then plotted as a function of mitochondrial respiratory rate (FIGURE 4). In FIGURE 4 we see that when either the log phosphorylation potential or the [ATP]/[ADP] ratio decreases, there is an increase in the rate of respiration. These results are quite similar to those reported from a number of laboratories. However, from the cellular standpoint there are problems with these data. TABLE 2 presents current estimates of the cytoplasmic phosphorylation potential and [ATP]/[ADP] ratio in heart, liver, muscle, and brain. The measured values were derived from the analysis of perchloric acid extracts of these tissues[60,61] while the calculated values were derived from en-

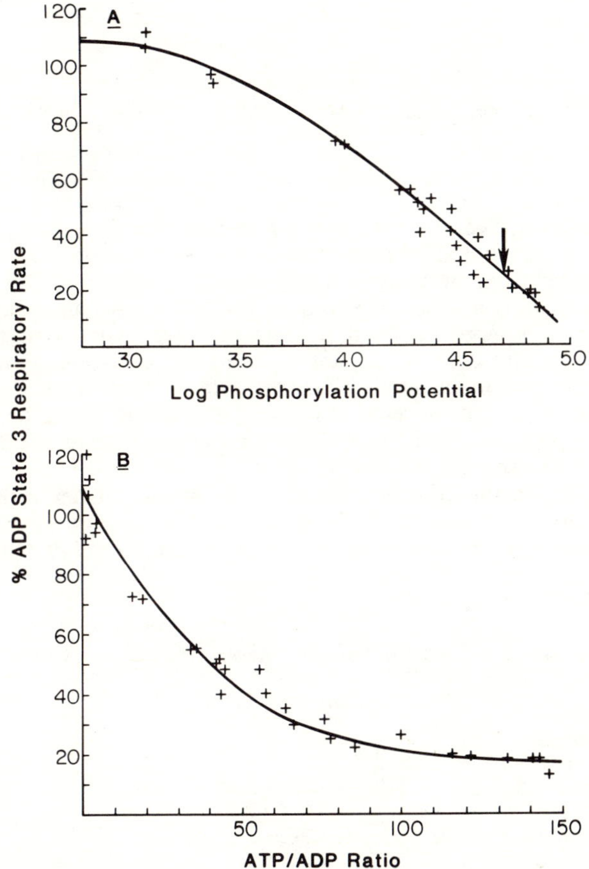

FIGURE 4. Correlation of respiratory rates to phosphorylation potentials (A) and [ATP]/[ADP] ratios (B), at constant [ATP] and variable hexokinase. Respiratory data were taken from FIGURE 3. Within 1 min after the addition of hexokinase (FIGURE 2), 2.5-ml samples were removed, acid-extracted, and assayed for [ATP], [ADP], and [P_i] as outlined under EXPERIMENTAL PROCEDURES. From these data, the indicated ratios were calculated. Phosphorylation potential data are expressed as M^{-1}. (From Jacobus *et al.*[33] With permission from *Journal of Biological Chemistry*.)

TABLE 2

COMPARISON OF ESTIMATED AND MEASURED VALUES FOR TISSUE POTENTIALS AND [ATP]/[ADP] RATIOS

Tissue	Measured	Calculated
Log cytoplasmic phosphorylation potential		
Brain	3.12*	4.48* 4.53†
Muscle	3.05*	4.43*
Liver	2.75*	4.21*
Heart	2.59‡	4.88¶
Cytoplasmic [ATP]/[ADP] ratios		
Brain	3.59*	82* 90†
Muscle	8.72	218*
Liver	2.65*	78*
Heart	5.70	534¶

Data derived from: * Veech *et al.*,[61] † Ackerman *et al.*,[62] ‡ Williamson *et al.*,[60] and ¶ Kohn *et al.*[63]

zyme equilibrium data,[60] ^{31}P NMR of the intact brain,[62] or from the computer analysis of heart adenine nucleotides.[63] In TABLE 2 we note that these new phosphorylation potentials, or [ATP]/[ADP] ratio values are at least ten-fold higher than those previously estimated from the analysis of acid-extracted tissues.[60,61] This results from the fact that the tissue content of ADP is quite low.[61–63] We presume that the reason for this is that *in vivo* a large fraction of the acid-extracted ADP was tightly bound to proteins. Tightly bound ADP would not be detectable by NMR, as well as being thermodynamically inconsequential. Thus, the data of TABLE 2 raise a significant new question: How do cells have rapid rates of respiration[2] when the cytoplasmic phosphorylation potentials and [ATP]/[ADP] ratios are high and thus fall in the presumed inhibitory range?

In order to resolve this apparent and important paradox, a new series of experiments were performed. In this protocol the content of hexokinase was held constant at an amount such that the maximum rate of respiration in the presence of 1 mM

TABLE 3

APPARENT K_m ADP VALUES FOR RESPIRATORY STIMULATION

	Constant ATP		Constant Enzyme
Rat liver mitochondria			
Direct ADP		22.6	
Agarose hexokinase	40.0		21.3
Yeast hexokinase	56.0		14.6
Skeletal muscle creatine kinase	26.2		12.5
Rat heart mitochondria			
Direct ADP		15	
Agarose hexokinase	24.2		19.5
Yeast hexokinase	13.9		6.8
Endogenous creatine kinase			2.8

All K_m values were derived from linear regression of double-reciprocal plots and are expressed as μM concentration values.

ATP was approximately 90% of the ADP State 3 rate. We assumed that under these conditions the rate of ADP production by hexokinase could never overdrive the capacity of electron transport and oxidative phosphorylation to rephosphorylate ADP to ATP. Therefore, these reactions could not be rate limiting. Under the conditions, it was possible to change the steady-state rates of respiration by varying the concentration of ATP initially added to the oxygraph chamber. In FIGURE 5 we see that as we increase the content of ATP from 5 μM to 10 mM, one progressively increases the rate of respiration in a manner similar to the results seen in FIGURE 3. Constant steady-state rates of respiration were easily achieved. However, when we extract samples and compare the phosphorylation potential data and the [ATP]/[ADP] ratios to the respiratory rates, a picture quite different from FIGURE 4 emerges. The data of FIGURE 6 correlates low rates of respiration to low phosphorylation potential or [ATP]/[ADP] ratio values, and as these parameters increase the rates of respiration also increases. The data of this FIGURE are thus strikingly opposite to the data of FIGURE 4. Therefore, when one compares the data of FIGURES 4 and 6, one is drawn to the conclusion that neither the extramitochondrial phosphorylation potentials nor the [ATP]/[ADP] ratios can be the absolute parameters that control the rates of respiration. If such were the case, then the data from the two protocols should be equivalent and this clearly is not the case. In fact, at a constant log phosphorylation potential value of 3.7 (FIGURE 6A), or an [ATP]/[ADP] ratio value of 10 (FIGURE 6B), the rates of respiration vary between 35% to 120% of State 3. Therefore, these data alone (FIGURE 5) show that there is little meaningful correlation between these metabolic parameters and the rates of mitochondrial respiration.

How do we reconcile this apparent discrepancy? The data of FIGURE 7 show the correlations between the concentrations of ADP and respiratory rates. In both in-

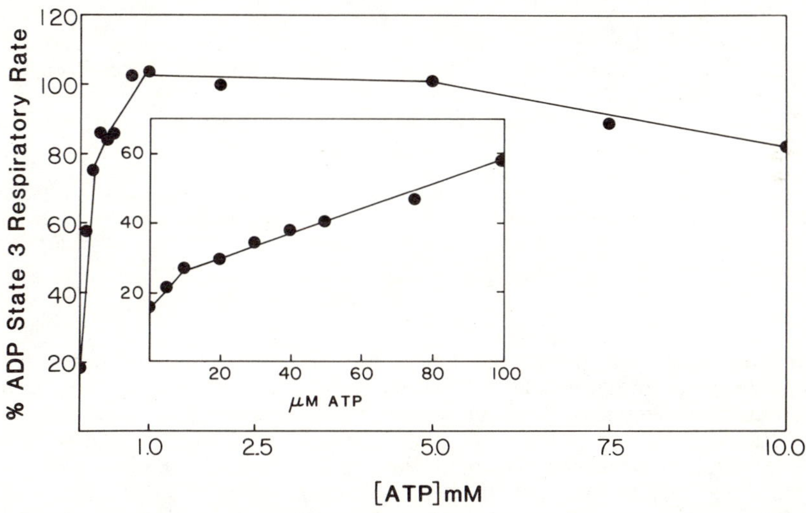

FIGURE 5. Titration curve for respiratory stimulation at constant hexokinase and variable [ATP]. The amount of hexokinase added was limited to that sufficient to stimulate respiration to approximately 90% ADP State 3. The indicated concentration of ATP was added first, then followed by the addition of hexokinase. The inset presents data at low concentrations of ATP, between zero and the first data point in the full figure. (From Jacobus *et al.*[33] With permission from *Journal of Biological Chemistry*.)

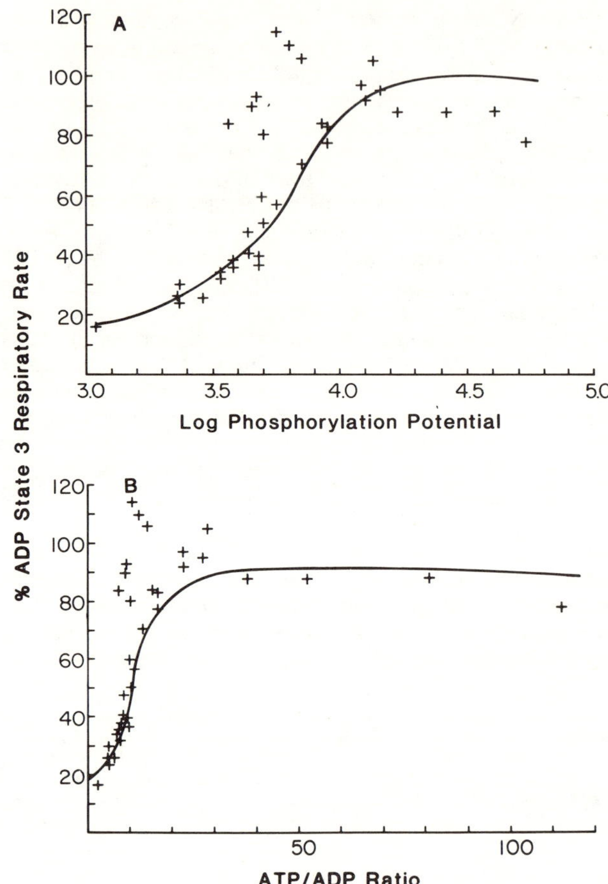

FIGURE 6. Correlation of respiratory rates to phosphorylation potentials (A) and [ATP]/[ADP] ratios (B), at constant hexokinase and variable [ATP]. Respiratory rates were taken from FIGURE 5 and samples were processed as described in FIGURE 4. (From Jacobus *et al.*[33] With permission from *Journal of Biological Chemistry*.)

stances respiratory rates increase as a function of ADP concentration. These data, in double-reciprocal format, can be used to estimate the apparent K_m ADP for respiratory stimulation (FIGURE 8). The values from this figure are 56 μM for the condition of constant ATP (1.0 mM), and 14.6 μM with constant hexokinase. These differences likely reflect the competitive inhibition of ADP binding to the adenine nucleotide translocase in the presence of the high background 1 mM ATP. The data of FIGURES 7 and 8 thus support the initial notion of Chance and Williams[36,37] that ADP availability was the parameter controlling the rates of cellular respiration.

In summary, we believe that the best interpretation of these data is that respiratory control is simply a function of the rate of transport of ADP into the mitochondrial matrix (FIGURES 7 and 8), mediated by the kinetic properties of the adenine nucleotide translocase.[57]

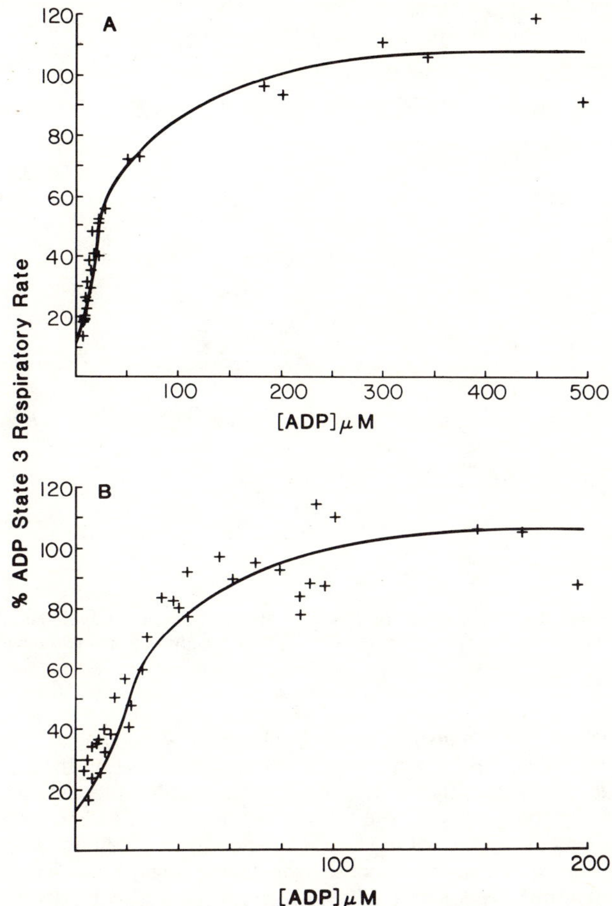

FIGURE 7. Respiratory rates as a function of [ADP]. (A) the condition of constant [ATP] and variable hexokinase (FIGURE 4). (B) the condition of constant hexokinase and variable [ATP] (FIGURE 6). (From Jacobus *et al.*[33] With permission from *Journal of Biological Chemistry*.)

Microcompartmentation of ADP Production

It is well appreciated that within the cell there exist a series of microcompartments. These can most easily be defined by observable diffusion barriers, i.e. intracellular organelles with limiting membranes. Examples are the mitochondrial matrix, the sarcoplasmic reticulum, and the volume encased by the lysosomal membranes. In short, these represent specific compartments clearly defined by unique membranes. It has also been suggested that within cells other forms of microcompartments may exist. These can be the result of diffusion barriers, enzyme-enzyme proximity, bound-water effects, and Nernst unstirred layers.[64] Each of these parameters can lead to the kinetic enhancement of an enzymatic reaction. Such effects have been described for natural, as well as artificial, multi-step enzyme complexes.[65–69] In a simi-

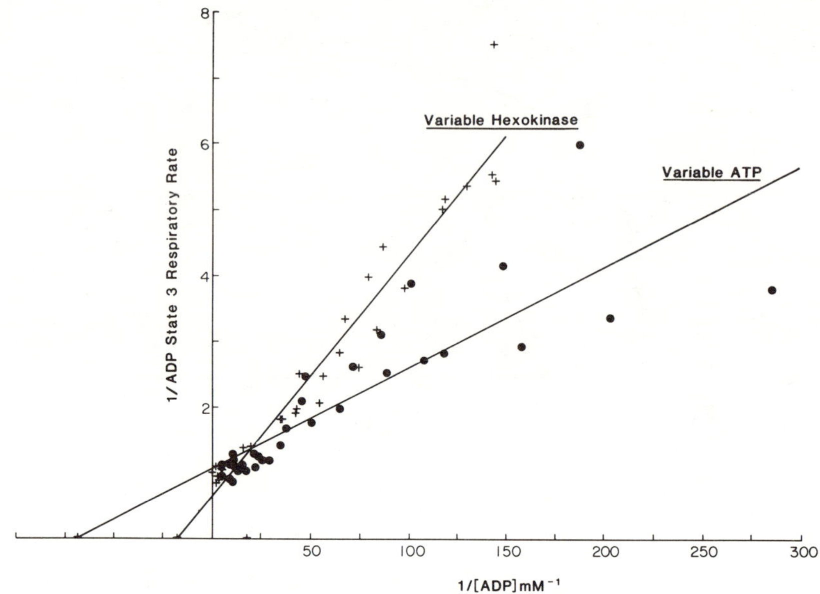

FIGURE 8. Double-reciprocal plots of 1/ADP State 3 respiratory rates versus 1/[ADP]. Data were derived from FIGURE 7. (From Jacobus *et al.*[33] With permission from *Journal of Biological Chemistry*.)

lar manner, such effects can also influence the kinetics of transmembrane metabolite transport. With respect to transport across the inner mitochondrial membrane by carrier systems, there are three relevant compartments: the extramitochondrial space, the outer membrane, and the intramembrane space or surface of the inner membrane (FIGURE 9). Thus, with respect to the control of cellular respiration, it is quite possible that the adenine nucleotide translocase might be sensitive not only to ADP availability, but the kinetics of translocation may also be influenced by the compartment in which the ADP was initially generated. If true, these three regions could therefore represent unique microcompartments within the cell.

To test this hypothesis of "vectorial" ADP production, three ADP-generating systems were employed (FIGURE 9). To simulate ADP generation in the cytoplasm, we used agarose-immobilized hexokinase (FIGURE 9A). Bound to a large agarose bead, it is improbable that the enzymes could come in close proximity to the translocase or interact with the hexokinase-binding protein of the outer membrane.[70] As a model of outer membrane ADP generation, we used yeast hexokinase (FIGURE 9B). It is well understood that hexokinase can bind to a specific hexokinase-binding protein localized in the outer membrane, and that the binding of the enzyme is regulated by various metabolites. This represents the second system. The endogenous heart mitochondrial isozyme of creatine kinase represents the third condition, an enzyme that generates ADP in close proximity to the translocase.[28,29] Intact rat liver and heart mitochondria were incubated in the presence of these enzymes, with ATP held constant at 1.0 mM and enzyme activity varied, or enzyme activity held constant and ATP titrated. In the cases of hexokinase, the concentration of glucose was 20 mM; with creatine kinase the creatine concentration was also 20 mM. These are saturating concentrations of phosphate acceptor. Otherwise, the reaction protocols were nearly

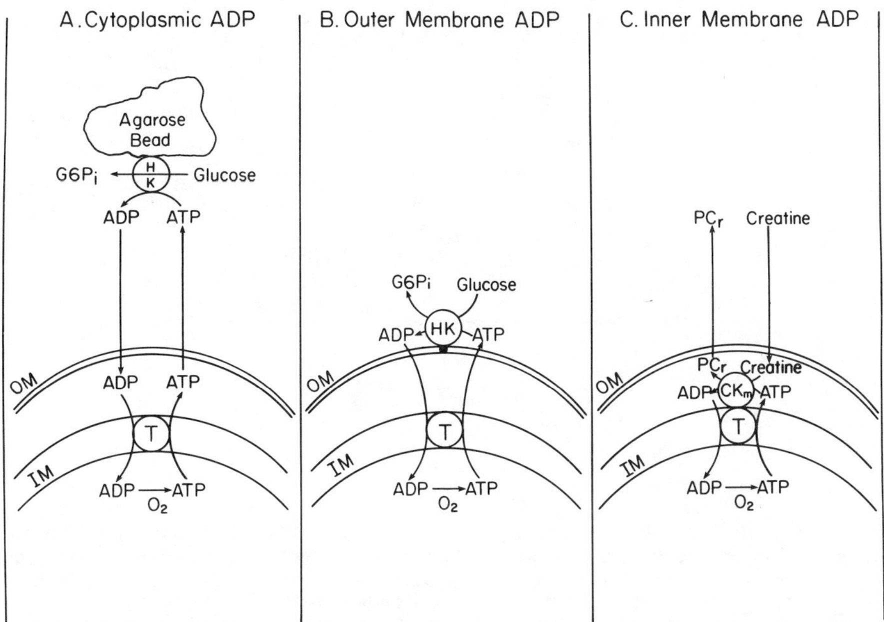

FIGURE 9. Model for compartmentalized ADP formation. (A) Simulation of cytoplasmic ADP formation using yeast hexokinase (HK) bound to large agarose beads. (B) Simulation of ADP generated near the surface of the outer membrane using soluble yeast hexokinase. (C) Simulation of ADP formation in the close proximity of the adenine nucleotide translocase (T) utilizing the endogenous mitochondrial isozyme of creatine kinase (CK_m).[25-29] Other abbreviations are: OM, outer mitochondrial membrane; IM, inner mitochondrial membrane; $G6P_i$, glucose 6-phosphate; and PCr, phosphocreatine.

equivalent to those used for FIGURES 7 and 8. After steady-state rates of respiration were achieved, samples were removed, extracted and assayed for [ADP]. These data were plotted in double-reciprocal form and estimates of K_m derived from the extrapolated abscissa intercepts. In all cases, the estimation of [ADP] was corrected for the small contribution by matrix ADP.[28]

In TABLE 3 we report the estimates of the apparent K_m ADP for these three systems, measured in sucrose oxygraph medium. Also reported are data derived from the direct addition of known concentrations of ADP. The important point that emerges from these studies is the fact that the apparent K_m for ADP translocation is markedly influenced by the location of ADP generation. The highest K_m values are observed for the direct addition studies, or those reactions with the immobilized hexokinase. The lowest K_m value is observed for the endogenous heart mitochondrial creatine kinase. These results are consistent with much previous data concerning the special relationship between mitochondrial creatine kinase and the adenine nucleotide translocase (FIGURE 10).[11,23-29] These data suggest that in heart muscle ADP generated by mitochondrial creatine kinase will have kinetic preference for translocation. Thus it appears that the rate of the forward creatine kinase reaction could be the dominant system regulating the rates of heart oxygen consumption.

The validity of the previous statement is further enhanced when one considers another aspect of ADP transport, namely its inhibition by ATP. When we compare

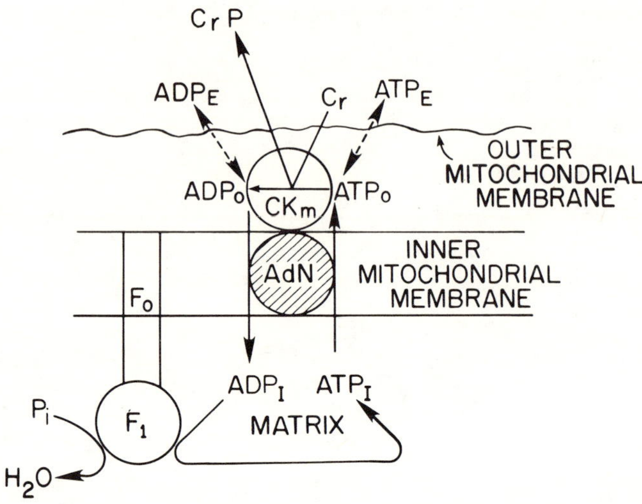

FIGURE 10. Model for the coupling of phosphocreatine synthesis to oxidative phosphorylation at the inner mitochondrial membrane. Abbreviations used are F_0F_I, proton-translocating ATPase; ADP_E and ATP_E, exogenous nucleotides; ADP_0 and ATP_0, nucleotides in proximity of CK_m-active site; ADP_I + ATP_I, matrix nucleotide pool; and Cr and CrP, creatine and creatine phosphate, respectively. (From Moreadith & Jacobus.[28] With permission from *Journal of Biological Chemistry*.)

our data to the reported kinetic properties of the translocase, some discrepancies become apparent. Pfaff and Klingenberg[71] first suggested that ATP interacts at the translocase in a manner competitive with ADP exchange. Souverijn *et al.*[72] confirmed these results and estimated that the K_I for ATP inhibition was in the 100 to 200 μM range. It must be noted that these later data were obtained at 0°C, with oxidative phosphorylation inhibited by oligomycin. When we substitute our data into the equation of Souverijn *et al.*,[72] using our value for the apparent K_m ADP, 14.6 μM, and their values for K_I ATP, 100 to 200 μM, we calculate that at 10 mM ATP respiration should only be occurring at a rate of 5% to 10% of the State 3 rate. This clearly contradicts the data of FIGURE 5, where respiration at 10 mM ATP is occurring at 80% of State 3. We have therefore redetermined the K_I for the ATP inhibition of respiration under phosphorylating conditions at 37°C (data not shown). These data suggest that the K_I ATP is not in the 100–200 μM range,[75] but is actually closer to 5 to 10 mM for liver mitochondria. This very low ability of ATP to inhibit State 3 respiration is additional evidence for why theories of respiratory control formulated on a simple [ATP]/[ADP] ratio are invalid. Such a ratio would only be valid if ATP is a strong, competitive inhibitor of ADP binding. Together with more recent data,[73] it appears that our knowledge of the kinetics and mechanistic properties of the adenine nucleotide translocase is incomplete. The differences become even more striking in heart mitochondria. In the system using agarose-immobilized hexokinase, the K_I value is estimated to be near 10 mM. Under the conditions where creatine kinase is generating the ADP, we estimate that the K_I for ATP inhibition is greater than 30 mM. This significant threefold difference means that changes in the cytoplasmic concentrations of ATP will have little influence on ADP transport during creatine kinase–stimulated respiration. This is another important piece of data underscoring the reason why mitochondrial creatine kinase would be the dominant reaction directly controlling O_2 consumption.

Summary

Three important points must be emphasized in summary. First is the idea that a cellular microcompartment need not be limited by a semi-permeable membrane. We recognize microcompartments in multi-enzyme complexes where substrates are covalently transported from subunit to subunit. An example of this is the lipoic acid moiety of the pyruvate dehydrogenase complex. However, to act as a kinetic microcompartment, covalent transfer is not an obligatory requirement. Proximity effects may be sufficient for substantial rate enhancement. Our data clearly show that the kinetics of ADP translocation are influenced by the site of ADP formation. We contend that this represents a newly recognized and important form of cellular microcompartmentation.

The second point is that we do not want our results misinterpreted as an overextension of the known data concerning tissue respiration. We believe that the primary parameter controlling heart mitochondrial oxygen consumption is the availability of ADP at the adenine nucleotide translocase. Our data show, however, that this is not a simple process. Secondary control is exerted by the localization of ADP formation, i.e. microcompartmentation. As a result of the kinetic data (Table 3), we conclude that the forward rate of mitochondrial creatine kinase is the preferential reaction controlling ADP delivery to the translocase. We are left, nonetheless, with questions concerning the secondary regulation of this enzyme *in vivo* by substrate (ATP and creatine) and inhibition by product (phosphocreatine). The nature of this control awaits further experimental data. Finally, the results are consistent with the creatine kinase energy transport hypothesis. Overall, the rate of tissue oxygen consumption reflects the metabolic activity of the organ, determined by the rate of ATP utilization (see right side of Figure 1). This results in the cytoplasmic production of ADP. In heart, this is coupled via the bound cytoplasmic isozymes of creatine kinase to the local rephosphorylation of ADP to ATP and the simultaneous production of creatine. Creatine then becomes an important regulatory signal to stimulate mitochondrial creatine kinase. In skeletal muscle such relationships have been described by Mahler.[74–76] The final link is the mitochondrial creatine kinase translocase activity, as described in this work.

The third important point is that we recognize the results presented in this paper address only the issues of the adenine nucleotide control of respiration. While our data are consistent with the view that the adenine nucleotide translocase rate limits oxidative phosphorylation,[57] it is also clear that other factors play important roles within the cell. These factors include oxygen,[47,48] the concentration and transport of phosphate,[41,50,57] substrate availability,[39,40,42] and matrix [ATP]/[ADP] ratios.[77,78] Defining the interactions of each of these factors under physiological conditions will provide a more complete picture of the cellular control mechanisms involved in the obligate regulation of oxidative phosphorylation. How changes in membrane structure might also effect this regulation represent another potential area for future investigation.

References

1. Neely, J. R. & H. E. Morgan. 1974. Annu. Rev. Physiol. **36**: 413–459.
2. Kobayaski, K. & J. R. Neely. 1979. Circ. Res. **44**: 166–175.
3. Gudbjarnason, S., P. Mathes & K. G. Ravens. 1970. J. Mol. Cell. Cardiol. **1**: 325–339.
4. Seraydarian, M. W. & B. C. Abbott. 1976. J. Mol. Cell. Cardiol. **8**: 741–746.
5. Saks, V. A., L. V. Rosenshtraukh, V. N. Smirnov & E. I. Chazon. 1978. Can. J. Physiol. Pharmacol. **56**: 691–706.

6. Jacobus, W. E. & J. S. Ingwall, Eds. 1980. Heart Creatine Kinase: The Integration of Isozymes for Energy Distribution. Williams & Wilkins Co. Baltimore.
7. Bessman, S. P. & P. J. Geiger. 1981. Science **211**: 448–452.
8. Bessman, S. P. & A. Fonyo. 1966. Biochem. Biophys. Res. Commun. **22**: 597–602.
9. Vial, C., C. Godinot & D. Gautheron. 1972. Biochimie **54**: 843–852.
10. Sobel, B. E., W. E. Shell & M. S. Klein. 1972. J. Mol. Cell. Cardiol. **4**: 367–380.
11. Jacobus, W. E. & A. L. Lehninger. 1973. J. Biol. Chem. **248**: 4803–4810.
12. Saks, V. A., G. B. Chernousova, Y. I. Voronkov, V. N. Smirnov & E. I. Chazov. 1974. Circ. Res. **34** and **35** (Suppl. III): 138–149.
13. Groose, R., E. Spitzer, V. V. Kupriyanov, V. A. Saks & K. R. H. Repke. 1980. Biochim. Biophys. Acta **603**: 142–156.
14. Levitsky, D. O., T. S. Levchenko, V. A. Saks, V. G. Sharov & V. N. Smirnov. 1978. Membr. Biochem. **2**: 81–96.
15. Bessman, S. P., W. C. T. Yang, P. J. Geiger & S. Erickson-Viitanen. 1980. Biochem. Biophys. Res. Commun. **96**: 1414–1420.
16. Saks, V. A., N. V. Lipina, V. G. Sharov, V. N. Smirnov, E. Chazon & R. Grosse. 1977. Biochim. Biophys. Acta **465**: 550–558.
17. Baskin, R. J. & D. W. Deamer. 1970. J. Biol. Chem. **245**: 1345–1347.
18. Farrell, E. C. & N. Baba. 1974. *In* Electron Microscopy of Enzymes: Principles & Methods. M. A. Hayet, Ed. **13**: 135–152. Van Nostrand & Reinhold. New York.
19. Sharov, V. G., V. A. Saks, V. N. Smirnov & E. Chazov. 1977. Biochim. Biophys. Acta **468**: 495–501.
20. Ottaway, J. H. 1967. Nature **215**: 521–552.
21. Scholte, H. R. 1973. Biochim. Biophys. Acta **305**: 413–427.
22. Saks, V. A., G. B. Chernousova, R. Vetter, V. N. Smirnov & E. I. Chazov. 1976. FEBS Lett. **62**: 293–296.
23. Saks, V. A., G. B. Chernousova, D. E. Gukovsky, V. N. Smirnov & E. I. Chazov. 1975. Eur. J. Biochem. **57**: 273–290.
24. Saks, V. A., N. V. Lipina, V. N. Smirnov & E. I. Chazov. 1976. Arch. Biochem. Biophys. **173**: 34–41.
25. Saks, V. A., V. V. Kupriyanov, G. V. Elizarova & W. E. Jacobus. 1980. J. Biol. Chem. **255**: 755–763.
26. Yang, W. C. T., P. J. Geiger, S. P. Bessman & B. Borrebaek. 1977. Biochem. Biophys. Res. Commun. **76**: 882–887.
27. Jacobus, W. E. & V. A. Saks. 1982. Arch. Biochem. Biophys. **219**: 167–178.
28. Moreadith, R. W. & W. E. Jacobus. 1982. J. Biol. Chem. **257**: 889–905.
29. Gellerrich, F. & V. A. Saks. 1982. Biochem. Biophys. Res. Commun. **105**: 1473–1481.
30. Vercesi, A., B. Reynafarje & A. L. Lehninger. 1978. J. Biol. Chem. **253**: 6379–6385.
31. Schneider, W. C. 1957. *In* Manometric Techniques. W. W. Umbreit, R. Burris & F. J. Stauffer, Eds.: 188–199. Burgess Publishing Co. Minneapolis.
32. Bradford, M. 1976. Anal. Biochem. **72**: 248–254.
33. Jacobus, W. E., R. W. Moreadith & K. M. Vandegaer. 1982. J. Biol. Chem. **257**: 2397–2402.
34. Lamprecht, W. & I. Trautschold. 1974. *In* Methods of Enzymatic Analysis. H. U. Bergmeyer, Ed. **4**: 2101–2110. Academic Press. New York.
35. Noda, L., S. Kuby & H. Lardy. 1955. Methods Enzymol. **2**: 605–610.
36. Chance, B. & G. R. Williams. 1955. J. Biol. Chem. **217**: 385–393.
37. Chance, B. & G. R. Williams. 1956. Adv. Enzymol. **17**: 65–134.
38. Klingenberg, M. 1961. Biochem. Z. **335**: 263–272.
39. Owen, C. S. & D. F. Wilson. 1974. Arch. Biochem. Biophys. **161**: 581–591.
40. Wilson, D. F., M. Stubbs, R. L. Veech, M. Erecinska & H. A. Krebs. 1974. Biochem. J. **140**: 57–64.
41. Holian, A., C. S. Owen & D. F. Wilson. 1977. Arch. Biochem. Biophys. **181**: 164–171.
42. Nishiki, K., M. Erecinska & D. F. Wilson. 1978. Am. J. Physiol. **234**: C73–C81.
43. Erecinska, M., D. F. Wilson & K. Nishiki. 1978. Am. J. Physiol. **234**: C82–C89.
44. Stubbs, M., P. V. Vignais & H. A. Krebs. 1978. Biochem. J. **172**: 333–342.
45. Erecinska, M., T. Kula & D. F. Wilson. 1978. FEBS Lett. **87**: 139–144.

46. VAN DER MEER, R., T. P. M. AKERBOOM, A. K. GROEN & J. M. TAGER. 1978. Eur. J. Biochem. **84**: 421–428.
47. WILSON, D. F., M. ERECINSKA, C. DROWN & I. A. SILVER. 1979. Arch. Biochem. Biophys. **195**: 485–493.
48. WILSON, D. F., C. S. OWEN & M. ERECINSKA. 1979. Arch. Biochem. Biophys. **195**: 494–504.
49. HOLIAN, A. & D. F. WILSON. 1980. Biochemistry **19**: 4213–4221.
50. SLATER, E. C., J. ROSING & A. MOL. 1973. Biochim. Biophys. Acta **292**: 543–553.
51. DAVIS, E. J. & L. LUMENG. 1975. J. Biol. Chem. **250**: 2275–2282.
52. KUSTER, U., R. BOHNENSACK & W. KUNZ. 1976. Biochim. Biophys. Acta **440**: 391–402.
53. DAVIS, E. J. & W. I. A. DAVIS-VAN THIENEN. 1978. Biochem. Biophys. Res. Commun. **83**: 1260–1266.
54. BOHNENSACK, R. & W. KUNZ. 1978. Acta Biol. Med. Ger. **37**: 97–112.
55. CHRISTIANSEN, E. N. & E. J. DAVIS. 1978. Biochim. Biophys. Acta **502**: 17–28.
56. REICHERT, M., H. SCHALLER, W. KUNZ & G. GERBER. 1978. Acta Biol. Med. Ger. **37**: 1167–1176.
57. LEMASTERS, J. J. & A. E. SOWERS. 1979. J. Biol. Chem. **254**: 1248–1251.
58. LETKO, G. & U. KUSTER. 1979. Acta Biol. Med. Ger. **38**: 1379–1385.
59. BOHNENSACK, R. 1981. Biochim. Biophys. Acta **634**: 203–218.
60. WILLIAMSON, J. R., S. W. SCHAFFER, C. FORD & B. SAFER. 1976. Circulation **53** (Suppl. I): I-3–I-14.
61. VEECH, R. L., J. W. R. LAWSON, N. W. CORNELL & H. A. KREBS. 1979. J. Biol. Chem. **254**: 6538–6547.
62. ACKERMAN, J. J. H., T. H. GROVE, G. G. WONG, D. G. GADIAN & G. K. RADDA. 1980. Nature **283**: 167–170.
63. KOHN, M. C., M. J. ACHS & D. GARFINKEL. 1977. Am. J. Physiol. **232**: R158–R163.
64. MOSBACH, K. 1978. *In* Microenvironments and Metabolic Compartmentation. P. A. Srere & R. W. Estabrook, Eds.: 381–400. Academic Press. New York.
65. DESIMONE, J. A. & S. R. CAPLAN. 1973. Biochemistry **12**: 3023–3039.
66. GOHEER, M. A., B. J. GOULD & D. V. PARKE. 1976. Biochem. J. **157**: 289–294.
67. NITISEWOJO, P. & H. O. HULTIN. 1976. Eur. J. Biochem. **67**: 87–94.
68. MALPIECE, Y., M. SHARAN, J. N. BARBOTIN, P. PERSONNE & D. THOMAS. 1980. J. Biol. Chem. **255**: 6883–6890.
69. ISHIKAWA, T., M. TAMURA & I. YAMAZAKI. 1980. J. Biol. Chem. **255**: 10764–10770.
70. FELGNER, P. L., J. L. MESSER & J. E. WILSON. 1979. J. Biol. Chem. **254**: 4946–4949.
71. PFAFF, E. & M. KLINGENBERG. 1968. Eur. J. Biochem. **6**: 66–79.
72. SOUVERIJN, J. H. M., L. A. HUISMAN, J. ROSING & A. KEMP, JR. 1973. Biochim. Biophys. Acta **305**: 185–198.
73. BARBOUR, R. L. & S. H. P. CHAN. 1981. J. Biol. Chem. **256**: 1940–1948.
74. MAHLER, M. 1979. J. Gen. Physiol. **71**: 533–557.
75. MAHLER, M. 1978. J. Gen. Physiol. **71**: 559–580.
76. MAHLER, M. 1980. *In* Heart Creatine Kinase: The integration of isozymes for energy distribution. W. E. Jacobus & J. S. Ingwall, Eds.: 92–108. Williams & Wilkins. Baltimore.
77. LETKO, G., U. KUSTER, J. DUSZYNSKI & W. KUNZ. 1980. Biochim. Biophys. Acta **595**: 196–203.
78. KUSTER, U., G. LETKO, W. KUNZ, J. DUSZYNSKY, K. BOGUCKA & L. WOJTCZAK. 1981. Biochim. Biophys. Acta **636**: 32–38.

LYSOPHOSPHATIDYLCHOLINE ACYLTRANSFERASE: PURIFICATION AND APPLICATIONS IN MEMBRANE STUDIES

David W. Deamer and Victor Gavino

Department of Zoology
University of California
Davis, California 95616

Introduction

In early studies, Lands[4] demonstrated an enzyme activity in rat liver microsomes that was capable of acylating lysophosphatidylcholine (LPC) to form phosphatidylcholine. This enzyme is now known as acyl coenzyme A:l-acyl-sn-glycero-3-phosphorylcholine acyltransferase (E.C.2.3.1.23) and will be abbreviated here as LPC acyltransferase. We became interested in LPC acyltransferase several years ago when it became apparent that the enzyme could be a valuable tool for investigations of membrane structure. Some potential applications are illustrated in Figure 1. These take advantage of the fact that the enzyme uses two soluble substrates (LPC and acyl CoA derivatives) to form a diacyl phospholipid (phosphatidylcholine) that is membranogenic.

In the first example, if a membrane contains endogenous acyltransferase activity, the enzyme can be used to synthesize lipid *de novo* (Figure 1.1 and 1.2). This result is potentially useful in synthesizing lipid probes for membrane studies, including radiolabeled, fluorescent, and spin-labeled probes. Figure 1.3 shows that acyltransferase can reconstitute membranes from solubilized components, a result that will be described in detail later. At lower concentrations, the substrates of the enzyme can be used to fuse membranes (Figure 1.4) and at higher concentrations sufficient to solubilized membranes, two or more membrane species can be hybridized (Figure 1.5).

Reconstitution Studies

In order to make further progress in understanding LPC acyltransferase, it will be necessary to purify and characterize the enzyme. This has been the primary thrust of both our own and others' recent work, and several groups have established procedures for partial purification of microsomal acyltransferase that utilize detergent solubilization, gel permeation chromatography, and sucrose density–gradient centrifugation.[3,5,6] Our laboratory has recently demonstrated that LPC acyltransferase activity is able to reconstitute microsomal membrane components solubilized with oleoyl CoA and LPC.[2] This work suggested a purification method, and will therefore be described in some detail.

The rationale was that substrates of the enzyme, for instance, LPC and oleoyl CoA, are single-chain amphiphiles and therefore have detergent-like properties. It follows that if they are added to a membrane system, partial or complete solubilization of the membranes will occur and mixed micelles containing a mixture of the original lipids, integral proteins, and the substrates will be produced. However, if the membranes originally contained endogenous LPC acyltransferase activity, the deter-

0077-8923/83/0414–0090$01.75/0

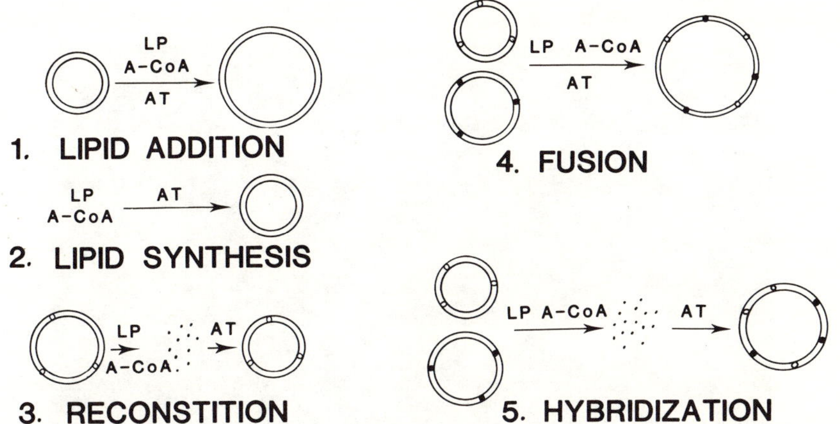

FIGURE 1. Potential applications of substrate detergent effects and acyltransferase activity in membrane studies. (See text.) LP, lysophosphatide; A-CoA, acyl coenzyme A; AT, acyltransferase.

gent effect will be degraded as phosphatidylcholine is synthesized, and lipid bilayers should form that incorporate integral membrane-protein components of the original microsomes.

To test this concept, rat liver microsomes were solubilized with additions of approximately 1–2 micromoles LPC and oleoyl CoA per mg protein, followed by incubation over a two-hour period. During this time, turbidity slowly increased, and it could be shown that most of the LPC was acylated to form phosphatidylcholine. Electron microscopy demonstrated the presence of membranous structures in the form of oligolamellar vesicles (FIGURE 2) and gel electrophoresis showed that most of the original protein components of the microsomes were present in the reconstituted membranes.

We conclude that acyltransferase held significant promise as a system for reconstituting certain membranes. Furthermore, milligram quantities of lipid were formed during this process, and it appeared likely that the enzyme, if isolated, could be valuable for synthesizing certain phospholipids that are difficult to synthesize chemically. For these reasons, the objective of our current research is to purify the enzyme as a first step toward realizing some of these goals. The results described above suggested a relatively simple isolation method that would involve the solubilization of membranes with sufficient LPC and acyl CoA so that approximately one protein molecule would be present in each mixed micelle. If that protein had acyltransferase activity, it would synthesize phospholipid. This would sufficiently alter the characteristics of the enzyme and surrounding lipid so that physical methods could be used to isolate only those lipid-protein complexes that contained the desired enzyme.

PURIFICATION OF LPC ACYLTRANSFERASE

Rat liver microsomes were isolated and washed with deoxycholate solutions as described by Hasegawa-Sasaki and Ohno.[3] This served to increase the specific activity of the enzyme about fivefold over the crude microsomes. The microsomes were

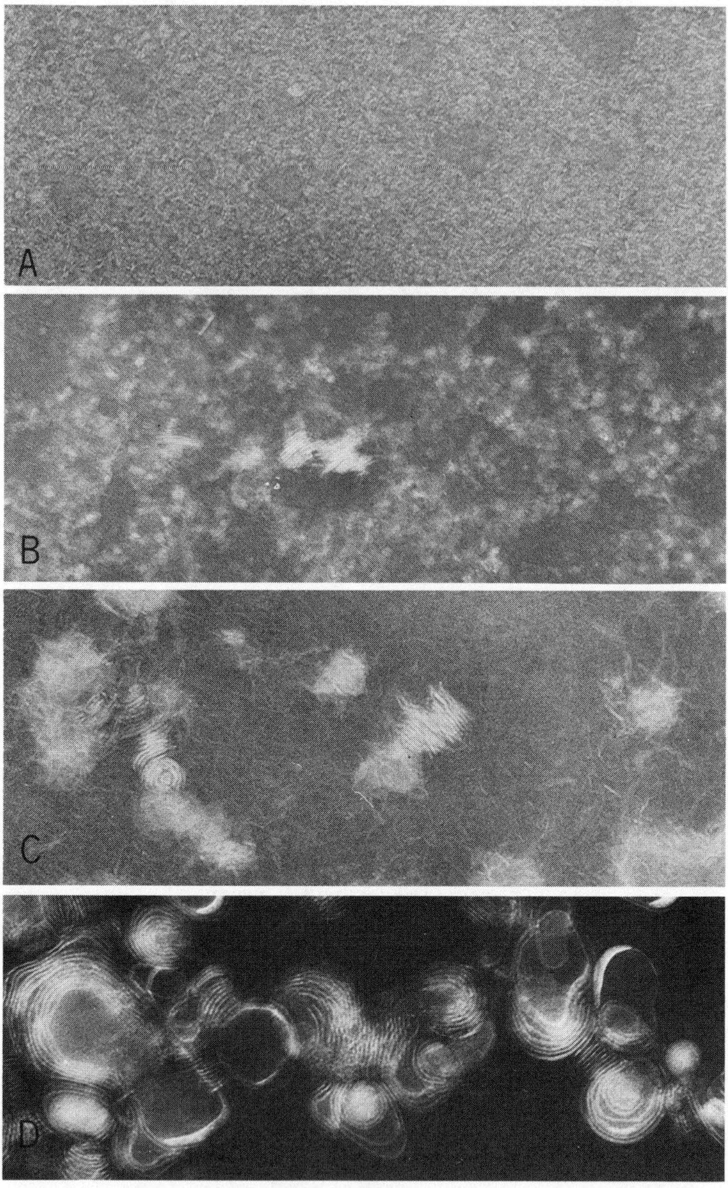

FIGURE 2. Negative stains of membranes formed by acyltransferase activity. (A) After solubilization by 2 μmoles LPC and oleoyl Co-A per mg protein, no membranous structures remained from the original liver microsomes. (B), (C), and (D) show the reconstitution process after 10, 30, and 120 minutes, respectively. ×87,000.

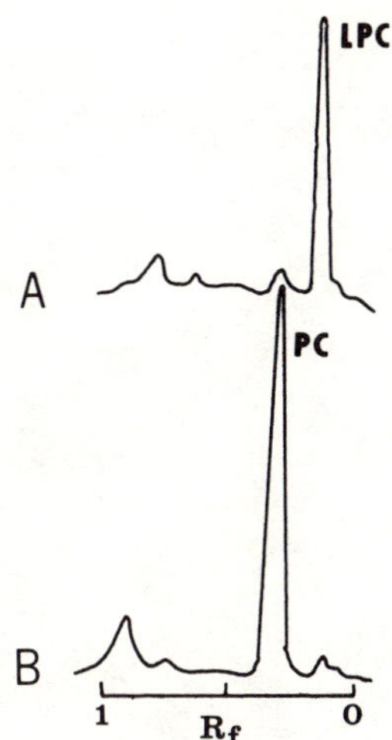

FIGURE 3. Lipid analysis of an acylation reaction mixture. Deoxycholate-washed microsomes (0.4 mg protein) were solubilized with 8 μmoles each of LPC and oleoyl CoA in 2.0 ml of 0.1 M Tris-HCl, pH 7.4. The reaction mixture was sampled at 5 min (A) and 60 min (B) and the lipids were extracted and analyzed by quantitative thin-layer chromatography using the Iatroscan instrument. At the start of the reaction, most of the phospholipid was represented by lysophosphatidylcholine (LPC) but after one-hour incubation, endogenous acyltransferase activity had converted the LPC to phosphatidylcholine (PC).

then solubilized with LPC at a lipid-protein ratio of about 10:1 by weight (20 μmoles LPC/mg protein) followed by the addition of an equimolar concentration of oleoyl CoA.

Acyltransferase activity was monitored by the chromogenic reaction of DTNB with free CoA, and was correlated with lipid synthesis by quantitative thin-layer chromatography on an Iatroscan TLC scanner. FIGURE 3A shows the lipid profile of the reaction mixture 5 min after the addition of oleoyl CoA. At this point, only a small amount of phosphatidylcholine (PC) was present relative to LPC. FIGURE 3B shows the lipid profiles of the reaction mixture after 60 min incubation. The specific activity of the microsomes used in this experiment was 330 nanomoles/min/mg and the substrate concentrations were 20 micromoles each of LPC and oleoyl CoA per mg microsomal protein. Under these conditions, the enzyme reaction should take at least 60 min to go to completion, and FIGURE 3B shows that the reaction was essentially complete after one hour.

FIGURE 4 shows the characteristic increase in light scattering that occurs during phosphatidylcholine synthesis under the conditions described for the acylation reaction in FIGURE 3. The increased light scattering reached a plateau after 45 to 50 minutes. Microsomes that were inactivated by heat showed no LPC acyltransferase activity by DTNB assay nor was increased light scattering observed upon incubation.

The acylation reaction mixture described in FIGURE 3 was mixed with an equal volume of glycerol and floated through a step gradient consisting of 30, 15, and 0

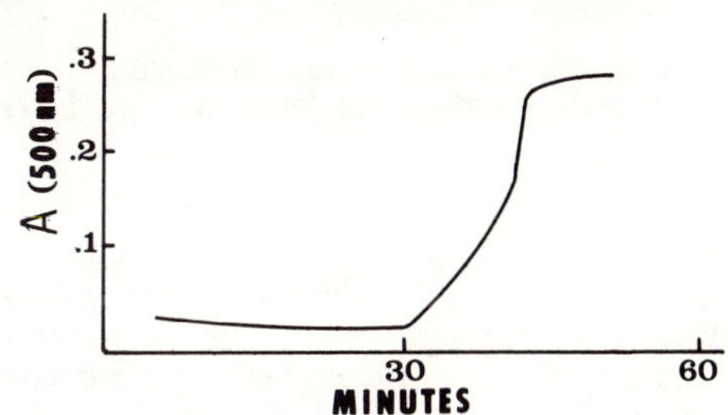

FIGURE 4. Increased turbidity during incubation reflects the formation of aggregates and membranous structures shown in FIGURE 2.

FIGURE 5. Freeze-fracture micrographs of liver microsomes (A) and lipid vesicles produced by acyltransferase activity (B). Note that the latter are relatively devoid of intramembrane particles. ×82,500.

percent glycerol. A turbid band collected at the interface of the 15 and 0 percent glycerol solutions. Turbid bands were not obtained from mixtures containing heat-inactivated deoxycholate-washed microsomes. Brown pellets always appeared at the bottom of the centrifuge tubes and were devoid of LPC acyltransferase activity. In two experiments, 9 and 11 percent of the total LPC acyltransferase activity was recovered in the tubid bands. Large amounts of membranous lipid were present, but

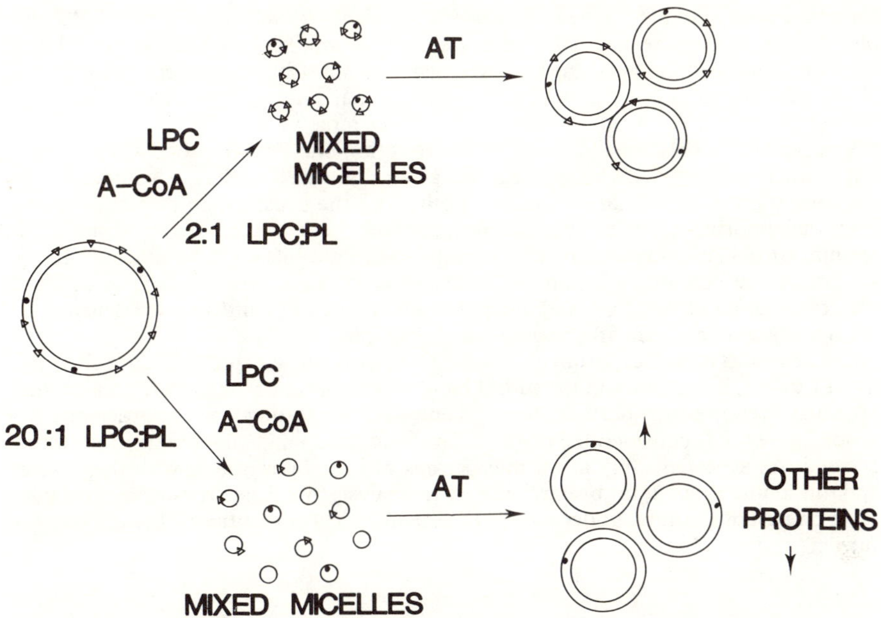

FIGURE 6. Summary of acyltransferase reconstitution and purification procedures. At lower concentrations of LPC and oleoyl CoA, mixed micelles are formed that presumably contain the added detergent-like substrates, together with the original membrane phospholipid and protein components. During acylation of the LPC these aggregate to form reconstituted membranes. At high LPC and oleoyl CoA concentrations (lower diagram) micelles are formed that ideally contain no more than one protein per micelle. If that protein has acyltransferase enzyme (dark circles) these micelles produce lipid-bilayer vesicles during acylation and may be separated from other excluded proteins by virtue of their decreased density.

the protein content of the bands was barely detectable by the micro-Lowry procedure.[1] Our best estimates for the protein content of the bands range from 1 to 4 micrograms with a specific activity of 9–10 micromoles/min/mg protein. This represents a 200-fold purification from the original microsomes.

The light-scattering increase shown in FIGURE 4 correlated with the appearance of vesicular structures by freeze-fracture electron microscopy. Typical vesicular structures of the original microsomes are shown in FIGURE 5A. These structures disappeared when the microsomes were solubilized with LPC and oleoyl CoA and were replaced by 10 nm micellar particles. After an incubation period of one hour, phosphatidylcholine synthesis resulted in the formation of membranous vesicular structures that ranged from 50 to 200 nanometers in diameter. The vesicles isolated by density-gradient centrifugation are shown in FIGURE 5B. These contain most of the enzyme activity and newly synthesized phosphatidylcholine. They are relatively devoid of intramembrane particles, as might be expected from the large excess of lipid bilayer present.

DISCUSSION

We have taken advantage of the fact that the product of acyltransferase activity can be phosphatidylcholine in the form of lipid bilayer vesicles. Our rationale was to solubilize deoxycholate-washed rat liver microsomes with a large excess of LPC and oleoyl CoA so that the original protein and lipid components would be diluted into mixed micelles composed of the two detergent-like substrates of the enzyme (FIGURE 6). At a 20:1 ratio of micromoles lipid/mg protein, and assuming an average molecular weight of 50,000 for microsomal proteins, the mole ratio of lipid to protein is in the range of 1,000:1. Ideally, each of the resulting micelles would contain at most one protein molecule. If that protein has acyltransferase activity, it would catalyze the formation of phosphatidylcholine, both from the substrate in that micelle and from neighboring micelles that do not contain the enzyme. When a sufficient amount of phosphatidylcholine has accumulated, lipid-bilayer vesicles would form. Assuming that little or no fusion with extraneous protein occurs, these vesicles could then be separated from the remaining protein by virtue of their decreased density, using density-gradient centrifugation as described earlier.

The results of our experiments generally support this scheme. The specific activity of LPC acyltransferase in the turbid band was an order of magnitude greater than previously reported for partially purified enzyme,[3] suggesting that the procedure described here holds considerable promise for the isolation and characterization of this enzyme. As expected, only a few micrograms of protein appeared with the vesicles and this amount has so far been refractory to analysis by gel electrophoresis. A large-scale preparation is now under way to collect sufficient quantities of the enzyme for further studies.

REFERENCES

1. BENSADOUN, A. & D. WEINSTEIN. 1976. Anal. Biochem. **70**: 241–250.
2. DEAMER, D. W. & D. E. BOATMAN. 1980. J. Cell Biol. **84**: 461–467.
3. HASEGAWA-SASAKI, H. & K. OHNO. 1980. Biochim. Biophys. Acta **617**: 205–217.
4. LANDS, W. E. M. 1960. J. Biol. Chem. **235**: 2233–2237.
5. WELTZIEN, H. U., G. RICHTER & E. FERBER. 1979. J. Biol. Chem. **254**: 3652–3657.
6. YAMASHITA, S., K. HOSAKA, Y. MIKI & S. NUMA. 1981. Methods Enzymol. **71**: 528–536.

THE AQUEOUS PORE IN THE RED CELL MEMBRANE: BAND 3 AS A CHANNEL FOR ANIONS, CATIONS, NONELECTROLYTES, AND WATER*

A. K. Solomon, B. Chasan,† James A. Dix,‡ Michael F. Lukacovic, Michael R. Toon, and A. S. Verkman

Biophysical Laboratory
Harvard Medical School
Boston, Massachusetts 02115

This article develops arguments for the existence of an aqueous pore in the red cell membrane as the principal route for passive flux of ions, water, and small nonelectrolytes and proposes a molecular model for the pore. In principle, such an aqueous pore would provide easy passage into and out of the cell for all solutes small enough to enter the channel. The red cell membrane, however, regulates the fluxes of cations and anions closely and discriminates carefully among other small solutes. These constraints have been incorporated into the model, which visualizes the channel and its associated regulatory system as governing passive transport of ions of either sign, as well as water and small nonelectrolytes into and out of the cell. The model, which was formulated to consolidate a number of observations already in the literature, has caused us to look for new interrelations between inhibitors specific to cation, anion, and nonelectrolyte transport. The results of these experiments, presented below, demonstrate that interrelations do exist and provide evidence that supports the view that a common aqueous channel provides primary access to the red cell cytoplasm.

Review of Evidence for the Aqueous Pore

In 1957, Paganelli and Solomon[1] measured water diffusion across the human red cell membrane and proposed that the membrane was traversed by an "equivalent pore" of 3.5 Å radius. The pore dimensions were established from the ratio of the osmotic permeability coefficient, P_f, measured by Sidel and Solomon,[2] to the diffusional permeability coefficient, P_d. Subsequently, Goldstein and Solomon[3] measured the reflection coefficient of a number of small hydrophilic solutes and found that these conformed to expectations for an aqueous pore whose dimensions could be computed from the equations governing steric hindrance put forward by Pappenheimer *et al.*[4] and Renkin.[5] Drawing primarily on these results, as well as a subsequent determination of P_d by Barton and Brown,[6] Solomon[7] calculated that $P_f/P_d = 3.4$ and concluded that the red cell equivalent pore had a radius of 4.2–4.6 Å.

Further support for the concept of an aqueous pore was provided by Vieira *et al.*[8] who measured the apparent activation energy for water diffusion into the human red cell and found it to be 6 kcal/mol, slightly greater than the 4.5 kcal/mol characteristic of water diffusion in bulk solution.[9] Vieira *et al.* found the activation energy

* Supported in part by National Institutes of Health grants 5R01 GM15692 and 2R01 HL14820 and National Science Foundation grant PCM-78-22577.

† Present address: Physics Department, Boston University, Boston, MA 02161.
‡ Present address: Department of Chemistry, SUNY, Binghamton, NY 13901.

0077-8923/83/0414-0097$01.75/0

for water diffusion into the dog red cell, whose equivalent pore radius was 6 Å, to be 4.5 kcal/mol, the same as in bulk solution, and ascribed the slightly larger value in human red cells to interactions between the water molecule and the hydrophilic walls of the smaller pore. Wang[9] measured water diffusion in bulk solution and computed the product $D_W \eta_W/T$ (in which D_W is the diffusion coefficient and η_W, the viscosity of bulk water), which contains all the temperature-dependent terms in the Stokes-Einstein coefficient for diffusion. Wang found this product to be independent of temperature for bulk water over the temperature range 5°C–35°C. An analogous product can be formed for diffusion into dog red cells by substituting P_d for D_W, as discussed by Solomon[10]; this product is also temperature independent ($P_d \eta_W/T$ = 1.58, 1.63, and 1.61 at 7°C, 22°C , and 37°C), supporting the conclusion that water behavior in small aqueous pores is similar to that in bulk solution.

Paganelli and Solomon's original estimate of the equivalent pore radius was based on the determination that $P_f/P_d > 1$. Following Mueller *et al.*'s[11] development of black lipid bilayers, it became possible to measure the P_f/P_d ratio in these bilayers, which contained no pores. The determination of Cass and Finkelstein[12] that the ratio was 1 in such bilayers provided further support for the conclusion that a ratio greater than 1 was, ordinarily, good evidence for the existence of an aqueous channel. The antibiotics, nystatin and amphotericin B, had been shown to produce aqueous channels in lipid bilayers and Holz and Finkelstein[13] measured the permeability characteristics of these channels to water and small nonelectrolytes. The antibiotic-induced pores were characterized by a P_f/P_d ratio of 3.3, virtually the same as the value of 3.4 in the human red cell. Furthermore, the steric hindrance of these induced pores is described accurately by the steric-hindrance equations given by Renkin[5] and can not be distinguished from that of the human red cell,[14] as shown in FIGURE 1 (top). DeKruijff and Demel[15] subsequently proposed a model for the amphotericin B pore consisting of a macromolecular assembly of eight molecules of amphotericin B and eight molecules of cholesterol forming a hollow cylinder, 20 Å long with an 8 Å inner diameter. Linear alignment of two of these macromolecular assemblies, one in each half of the bilayers, provides an unobstructed pore of 8 Å diameter running the entire length of the membrane. The inside of the pore is hydrophilic because the hydroxyl groups of the amphotericin B molecules form a continuous path along the inside of the pore and provide a hydrophilic environment for the water and solutes that traverse the pore; the conjugated double-bond system on the outside of the antibiotic molecules provides a lipophilic backbone that facilitates interaction with the bilayer lipids. An 8 Å pore would just exclude glucose, as is the case in both the human red cell and the pores induced in cholesterol-containing lipid bilayers.

This model has important implications concerning the red cell aqueous channel. It suggests that a uniform right cylindrical pore of 8 Å diameter offers steric restraints that match those in the human red cell. It does not mean that the aqueous channel in the red cell would have to be a right cylinder of 8 Å diameter, but it implies that the red cell pore must have properties equivalent to such a cylinder. Thus, small excursions about the mean diameter would be permitted, but abrupt departures from the mean, such as a constriction with a very much smaller aperture, would not conform to the observed steric hindrance.

Macey and Farmer[16] found that the sulfhydryl reagent, *p*-chloromercuribenzene sulfonate (pCMBS) inhibited water entrance through an aqueous channel. It took the reagent about 10–20 min to reach maximum effectiveness, at which time it inhibited osmotic water permeability by about 90% and diffusional permeability by about 50%, reducing the P_f/P_d ratio to 1; Macey and Farmer concluded that the effect of the reagent was to close the channel. We have used the data of Macey *et al.*[17] to compute that K_I for pCMBS inhibition = 0.2 mM and we have also confirmed this result

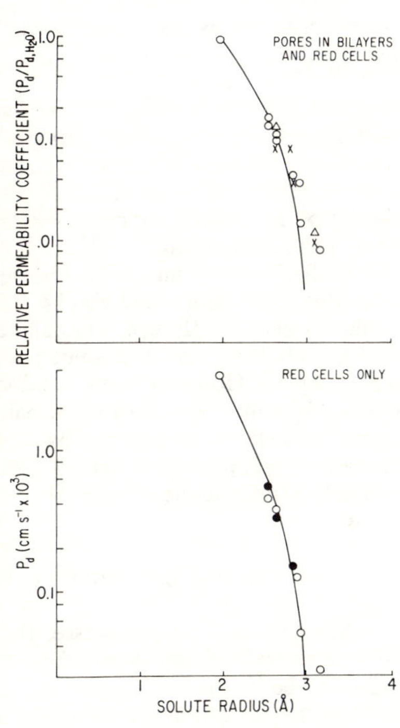

FIGURE 1. Comparison of steric hindrance of red cell pores with that of antibiotic-induced pores in lipid bilayers. The relative permeability coefficients for red cells have been computed from the data of Sha'afi *et al.*[104] and Savitz and Solomon[105] and those for lipid bilayers treated with nystatin and amphotericin B from that of Holz and Finkelstein.[13] The solute radii were computed from molar volumes based on crystalline densities. The lines in the figures were obtained by nonlinear least squares using the restricted diffusion equations of Renkin[5] and are displayed on logarithmic scales in order to make it easier to see the fit in the lower decades. Though the solute radius for water of 1.9 Å, determined from its molar volume, differs from the measured radius of 1.5 Å[106] and the best-fit pore radius of 3.1 Å differs from the value of 4.3–4.5 Å determined by other methods, the fit of the computed curve to the data is very good. Consequently, we have taken the curve as an effective empirical measure of hindrance to diffusion through aqueous channels that is operationally accurate over more than a decade of permeabilities. We have used these parameters, based on the curve in the bottom figure, to compute the hindrance that the aqueous channel offers to water diffusion. (Top) Comparison of permeability coefficients (relative to water) in red cell aqueous pores (○) with those induced in lipid bilayers by amphotericin B (×) and nystatin (△). Least squares radius = 3.04 Å. (Bottom) Fit of empirical curve to red cell permeability coefficients determined both by the minimum method described by Sha'afi *et al.*[104] (○) and by radioactive tracers as measured by Savitz and Solomon[105] (●). Least squares radius = 3.09 Å.

experimentally. Sha'afi and his colleagues[18-20] made a comparative study of the effect of many sulfhydryl reagents on water transport. They labeled the membrane proteins with a specific sulfhydryl reagent, [^{14}C]5,5′dithiobis-(2-nitrobenzoic acid) [^{14}C]DTNB), and reported that DTNB bound to the membrane sulfhydryls and also inhibited water transport. When the proteins were separated on a polyacrylamide gel,[21] Brown *et al.*[18] found that almost all the [^{14}C]DTNB migrated with the major red cell membrane protein, known as band 3,[22] which led them to conclude that the aqueous channel was associated with band 3.

Freeze-etch experiments of Pinto da Silva[23] and Pinto da Silva and Nicolson[24] provide further support for the view that the aqueous channel is associated with band 3. Pinto da Silva[23] observed that water sublimed from the membrane in regions underneath the "membrane intercalated particles," which are a characteristic feature of freeze-etch electron micrographs of human red cell ghosts. He interpreted this finding as evidence that the water initially below the particles had escaped through aqueous passages associated with these particles. Subsequently Pinto da Silva and Nicolson[24] used ferritin-labeled concanavalin A to identify band 3 as a component in the "membrane intercalated particles."

Band 3 is an integral membrane protein of about 95,000 daltons, which spans the membrane and is present in about 5.5×10^5 copies of a non-covalent dimer per cell.[25] It comprises some 25% of the membrane protein[26] and contains at least five cysteine residues.[27] In order to see if band 3 can account for the water flux, we have computed the number of aqueous pores of 4.5 Å radius that would be required to account for the measured diffusion coefficient of water across the membrane. For this purpose, we have used the P_{d,H_2O} of 4.7×10^{-3} cm sec^{-1} given by Barton and Brown[‡6]; the water diffusion coefficient, D_{H_2O},[9] of 2.3×10^{-5} cm^2 sec^{-1}; the red cell area, A_{mem},[33,34] of 1.35×10^{-6} cm^2; a pore radius, r_{pore}, of 4.5 Å; and length (ΔX) of 40 Å together with an estimate of steric hindrance ($A_{sd}/A_p = 0.064$ in the notation of Solomon[7]) determined from the parameters given in FIGURE 1(bottom). This computation[§] leads to a figure of 2.7×10^5 pores, which is in reasonable agreement with the 5.5×10^5 band 3 dimers in the membrane, bearing in mind the approximate nature of the calculation.

Only two other integral membrane proteins are present in a comparable number of copies, glycophorin and band 4.5. Though there are about 5×10^5 copies[35] of glycophorin per cell, glycophorin has been sequenced and found to contain no sulfhydryl groups,[36] so that it can not react with pCMBS and thus does not contain the aqueous pore. However, band 4.5 does contain SH groups. There are two recent estimates of the number of copies of band 4.5, 1.3×10^5 copies per cell[37] and 5.5×10^5 copies per cell[38] as determined by binding of cytochalasin B. Since there are 10–50% as many copies per cell of band 4.5 as of band 3, band 4.5 can not be excluded on numerical grounds alone.

Do Water and Nonelectrolytes Enter by a Lipid Pathway?

Macey *et al.*[17] have suggested that the red cell lipids provide the pathway for the 50% of water diffusion remaining after pCMBS treatment. However, the H_2O permeability coefficient of cholesterol-containing lipid bilayers (cholesterol comprises 40% of the red cell membrane lipid on a mole per mole basis) is too low, as can be seen from the following computation. The value of P_d given by Macey *et al.*[17] for pCMBS-treated red cells is 2.3×10^{-3} cm sec^{-1} at 25°C. If the diffusion took place through the lipids, which occupy 43% of the red cell area,[26,39] $P_{d,lipid}$ would have to be 5.3×10^{-3} cm sec^{-1}, in agreement with the value of 5.6×10^{-3} cm sec^{-1} computed similarly from the determinations of Conlon and Outhred[28] at 20°C. The highest

‡ There are a number of estimates of P_{d,H_2O} in the literature. We have chosen the one of Barton and Brown[6] since it agrees so well with that of Vieira *et al.*[8] and with measurements made by Barton and Brown in the original apparatus of Paganelli and Solomon. More recent experiments have been carried out with proton NMR[28-30] and ^{17}O NMR[31] methods, as well as one additional determination by the ^3H flow method.[32] The results are compared most easily in terms of the characteristic time constant for the diffusion process since this is independent of the cell volume, area, and water content, whose values have been revised several times since the original Paganelli and Solomon measurement made in 1957. Our present computation of diffusion time constants shows that the results fall primarily into two categories. Most of the proton NMR results have time constants of the order of 11–14 msec (after conversion to 20° C), which agree quite well with the 11.2 msec of Barton and Brown (11.9 msec at 20° C) used in the computation in the text. The values of Brahm[32] and the ^{17}O NMR value[31] are in the range of 19 to 22 msec (at 20° C). If these larger values were used in our computation, the number of pores required to carry the observed diffusion stream would drop by about 50%.

§The number of pores, n, is given by:

$$n = \frac{P_{d,H_2O}\, A_{mem}\, \Delta X}{(A_{sd}/A_p)\, \pi\, r_{pore}^2\, D_{H_2O}}.$$

value of P_d for cholesterol-containing phosphatidylcholine bilayers at 25°C is less than half of this value: those of Fettiplace[40] decrease from a maximum of 3.0×10^{-3} cm sec^{-1} to 1.4×10^{-3} cm sec^{-1} at maximal cholesterol concentration; that of Finkelstein[41] is 0.57×10^{-3} cm sec^{-1}.

Dix and Solomon[42] have investigated the lipid-permeation path by perturbing red cell membranes with the lipophilic anesthetic, halothane, which partitions into extracted red cell lipids and increases water permeability in red cell liposomes by 30% (20 mM halothane). However, 20 mM halothane has no effect on water permeability, either in intact cells or in cells maximally inhibited with pCMBS. These experiments suggest either that there is no significant water pathway through intact red cell lipids, or that the lipid structure in the intact cell differs significantly from liposome structure.

The irregularly shaped pCMBS molecule almost certainly does not close the aqueous channel by blocking it, like a cork in a bottle. Instead, as will be discussed below, it probably causes a conformational change in band 3, to which it binds chemically. If the conformational change and the attachment of the pCMBS molecule to band 3 occlude the channel significantly, water could still diffuse around the obstructions, though the diffusion coefficient would be much smaller than in the unimpeded channel, as Macey *et al.*[17] observed. In an unimpeded channel, P_f is larger than P_d because water-water friction in the flowing stream is much smaller than water-pore wall friction. However, when water flux is maximally inhibited by pCMBS, P_f/P_d is reduced to 1.18 ± 0.26.[17] The insertion of obstructions in such a narrow channel increases the fraction of collisions with the pore wall so that, in the limit, P_f approaches P_d and osmotic flow can no longer be differentiated from diffusional flow.[43,44]

Macey *et al.*[17] found that the apparent activation energy for water diffusion across the red cell membrane rises to 11.5 kcal/mol after pCMBS treatment, which agrees well with the 12 kcal/mole that Haran and Shporer[45] found for water diffusion across phosphatidylcholine-cholesterol vesicles and thus is consistent with diffusion through the membrane lipids. Alternatively, according to our model, this increased activation energy is a consequence of the necessity of bonds between H_2O and hydrogen-bonding sites in the wall of the aqueous pore, which become of increasing importance as the route of a permeant molecule around the pCMBS becomes more tortuous. Thus, Gary-Bobo and Solomon[46] found that the apparent activation energy for water diffusion through a hydrogen-bonding porous cellulose acetate membrane rose from the bulk solution value of 4.8 kcal/mol to 8.1 kcal/mol in a porous membrane with a P_f/P_d value of 3.2, not far different from that of the normal red cell membrane.

Macey *et al.*[17] showed that nonelectrolytes can still enter the red cell after 0.4 mM pCMBS treatment. At this concentration urea and water permeability coefficients are about 25% of the control values; glycerol permeability is inhibited by 50% and there is no effect on ethylene glycol. We attribute these differences among small hydrophilic nonelectrolytes to the result of differences in solute hydrogen-bonding properties in an aqueous pore whose structure has been distorted by pCMBS interactions with a band 3 sulfhydryl.[47] It is unlikely that the remaining 25% of the urea permeability, after pCMBS treatment, goes through the lipids. The permeability coefficient for urea transport across phosphatidylcholine vesicles is only about 1% of the urea permeability coefficient in whole red cells[48] and even lower in phosphatidylcholine-cholesterol bilayers,[41,49] which shows that urea is even less likely than water to enter the red cell by dissolving in the lipids. These observations do not rule out the existence of alternate specialized transport mechanisms for nonelectrolytes, but they strongly suggest that the lipids do not provide that pathway.

Locus of Sulfhydryl Group Controlling Water Permeability

Rao[27] and Rao and Reithmeier[50] has shown that the five SH groups on band 3 that react with N-ethylmaleimide (NEM) are in the cytoplasm of the cell. Three of the sulfhydryls are in the 43 kd cytoplasmic N-terminal segment of the protein and the other two are located in the 35 kd C-terminal segment as shown in Figure 2. All five SH groups can be protected by pCMBS against NEM and vice versa. Since pCMBS is an anion, it enters the red cell through the anion-exchange system, which apparently provides the only route by which pCMBS can penetrate to the cytoplasm. Although Knauf and Rothstein[51] suggested that pCMBS could also enter red cells through a channel that was not blocked by anion-transport inhibitors, the experiments of Rao[52] make this interpretation unlikely. She used an anion-transport inhibitor that greatly decreased the rate of pCMBS entrance into resealed ghosts and found that the half-time of the pCMBS reaction with the three SH groups on the 43 kd segment increased from 40 min to 190 min. Therefore, if any alternate route to the cytoplasm exists its half-time must be 190 min or greater, far longer than the 10–20 min incubation period for the pCMBS effect on water permeability.

In order to determine whether any of the five SH groups on band 3 was concerned with the pCMBS effect on water transport, we used the anion transport inhibitor, 4,4′-diisothiocyano-2,2′-disulfonic stilbene (DIDS) (100 μM) to covalently react with band 3 and block pCMBS access to the intracellular SH groups. When anion exchange was inhibited by 98% (in one experiment, typical of a second with 4,4′-dibenzamido-2,2′-disulfonic stilbene (DBDS) instead of DIDS), the ratio of P_f in treated cells (pCMBS + DBDS) to that in control cells (pCMBS alone) was 0.9 ± 0.1, consistent with unity, thus showing that the five SH groups inside the cell were not involved in the inhibition of water permeability. We confirmed this finding by pretreating the cells with NEM in a concentration large enough to interact with essentially all the SH groups on the membrane (12 mM, 1 hr, 37°C). After 25 min of pCMBS treatment, we found that water flux had been inhibited by $80 \pm 10\%$ in the controls (pCMBS alone) and $84 \pm 3\%$ in the NEM-treated cells (pCMBS + NEM) in one experiment, typical of two. This indicates that, if the effect of pCMBS is due to a sulfhydryl reaction, the SH group responsible does not react with NEM.

Steck *et al.*[53] and Ramjeesingh *et al.*[54] have reported an additional, sixth SH group in band 3. The studies of Ramjeesingh *et al.* were devoted to a 15 kd sequence contained in the membrane-bound 17 kd fragment that lies between the trypsin and chymotrypsin cleavage sites and contains the binding site for DIDS. Since Rao and Reithmeier[50] showed that there were no NEM-reactive groups in the 17 kd sequence, that sixth cysteine does not react with NEM; we will denote it as the cryptic sixth SH group. The schematic in Figure 2 shows an assumed disposition of band 3 in the red cell membrane, in which we have denoted the 17 kd sequence as the membrane-transport sequence because it contains both the DIDS binding site and the cryptic sixth SH group. As Jennings and Passow[55,56] and Rothstein and colleagues[57–59] have shown, portions of the 35 kd C-terminal fragment are also essential for anion transport; we have denoted this sequence as the membrane loop. As Figure 2 shows, we have considered the cryptic sixth SH group to be extracellular to the stilbene inhibition site. The argument to support this assignment is based on our observation that stilbene inhibitors do not alter the pCBMS effect on water transport, whereas they prevent access of pCMBS to the intracellular SH groups.

Vansteveninck *et al.*[60] reported that about 1–1.5% of the SH groups in the red cell membrane were on the outside of the cell and Rao[52] estimated that extracellular pCMBS sites on resealed red cell ghosts amounted to $1\text{–}2 \times 10^6$ sites/ghost. This class of sites was immediately accessible from the outside of the cell and binding was

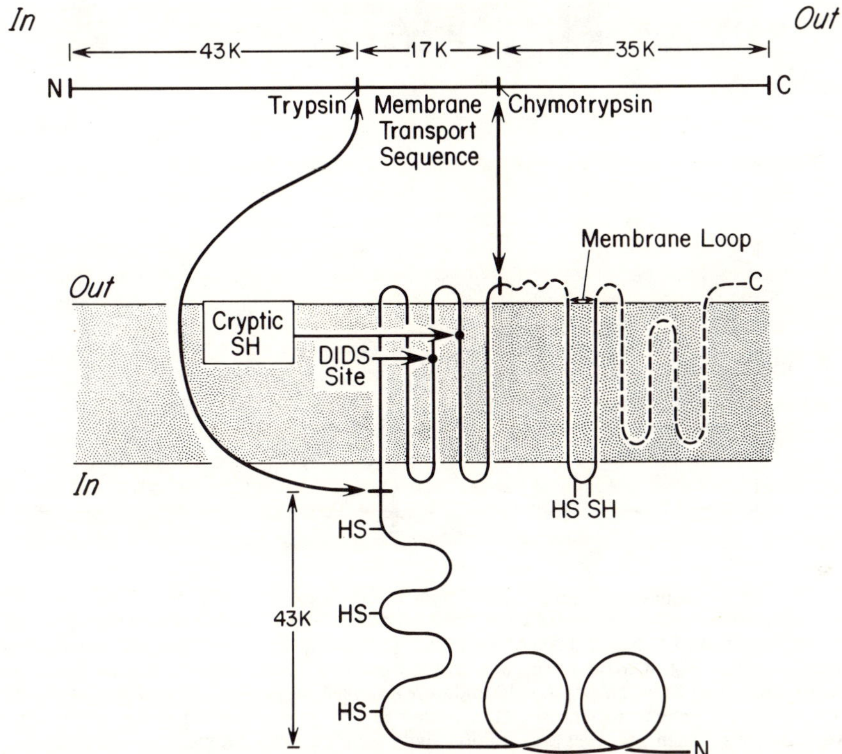

FIGURE 2. Schematic showing a possible disposition of band 3 in the red cell membrane. The location of the three intracellular SH groups on the 43 kd segment and the two intracellular SH groups on the 35 kd C-terminal fragment are taken from Rao.[52] However the number of loops in both of these fragments is purely arbitrary, as is the relative placement of the SH groups along the fragments. In the case of the 35 kd fragment, it is known that the 8 kd membrane loop crosses the membrane but it is not known how often the remainder of the 35 kd fragment traverses the membrane nor whether the C-terminal is extracellular or within the membrane.

not affected by inhibition of anion transport, so these sites presumably included the cryptic sixth SH group as well as the external SH groups[37] on band 4.5.

In order to determine whether the cryptic sixth SH group is actually located on band 3, we have examined the localization of the pCMBS site in NEM-treated resealed ghosts by polyacrylamide gel electrophoresis of the red cell membrane proteins using [^{203}Hg]pCMBS as shown in FIGURE 3. As the figure shows, the predominant peak is the one on band 3 and there is another peak at the end of the gel, beyond the tracking-dye position, which we have tentatively attributed to membrane lipids. FIGURE 3 confirms the existence of a sulfhydryl on band 3 that does not react with NEM and also shows that this SH group can react with pCMBS.

There is also a smaller peak at band 4.5 which, taking account of the lower molecular weight of band 4.5, could amount to some 30% of the number of sites on band 3. Considerable evidence has been presented indicating that the function of band 4.5 is to transport glucose and that an extracellular SH group is required for

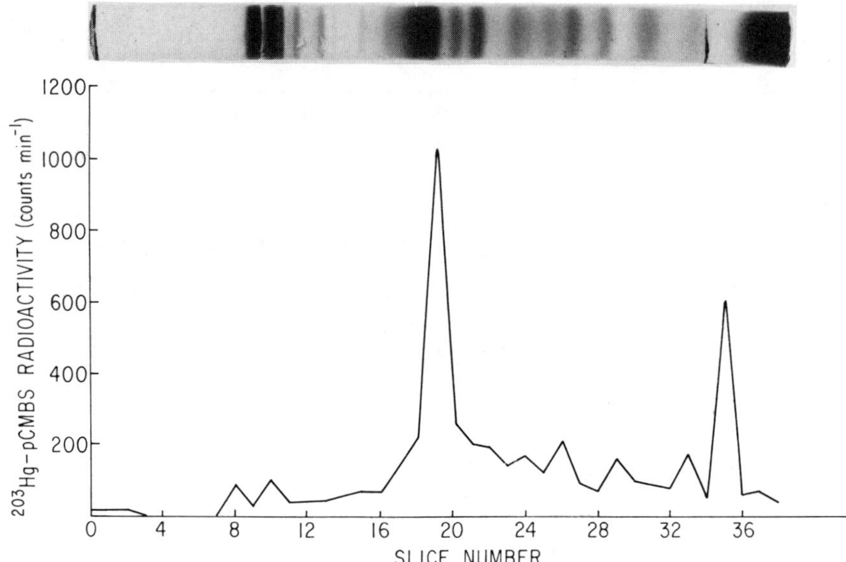

FIGURE 3. Staining and labeling profiles of red cell membrane proteins isolated from NEM-treated, resealed ghosts incubated with [^{203}Hg]pCMBS. Ghosts were treated with 2 mM NEM in PBS buffer (0.15 M NaCl; 0.005 M Na_2HPO_4, pH 7.5) for 1 hour at 25°C and then incubated with 0.1 mM [^{203}Hg]pCMBS in PBS buffer, for 2 min at 0°C. Ghosts were washed 3 × in PBS buffer containing 2 mM NEM (20 vols) and were analyzed for protein and radioactivity by SDS-polyacrylamide gel electrophoresis. Gels contained 5% polyacrylamide and 50 μg of protein were loaded on each gel. Gels with radioactivity were cut into 2 mm slices and counted.

this function.[61] Treatment with either pCMBS or NEM inhibits net glucose transport by 50 to 60% and it is not known whether there is a cryptic SH group on band 4.5 that exercises a physiological function.

Further evidence bearing on the interaction of pCMBS with band 3 comes from tryptophan fluorescence studies. Steck *et al.*[53] have found approximately 10 tryptophan residues in band 3, which Kleinfeld *et al.*[62] have shown to be primarily distributed in two arrays, one group extending about 10 Å into the cytoplasm and a larger group located near the outer surface of the membrane. Lukacovic *et al.*[63] have reported that pCMBS quenches tryptophan fluorescence (excitation 290 nm; emission 330 nm) in NEM-treated red cell ghosts, as shown in FIGURE 4 (top). There is an initial rapid quenching, not shown in FIGURE 4, that is too fast for us to resolve, followed by at least one slow component with a half-time of the order of 1–5 min. As FIGURE 4 (top) shows, similar effects are observed in reconstituted band 3 vesicles. These vesicles have been prepared from band 3 purified by the method of Lukacovic *et al.*[64] and contain no band 4.5. Since these band 3 vesicles have not been treated with NEM, it is possible that the other five SH groups may also have contributed to the fluorescence change, although the similarity of the time course to that of NEM-treated ghosts makes this unlikely. FIGURE 4 (bottom) shows the equilibrium fluorescence intensity of the tryptophan residues in reconstituted band 3 vesicles as a function of pCMBS concentration. We can conclude from FIGURE 4 that pCMBS interacts with the tryptophan residues in band 3 with a time course of the order of 1–5 min and a half point on the order of 0.1 to 0.3 mM pCMBS.

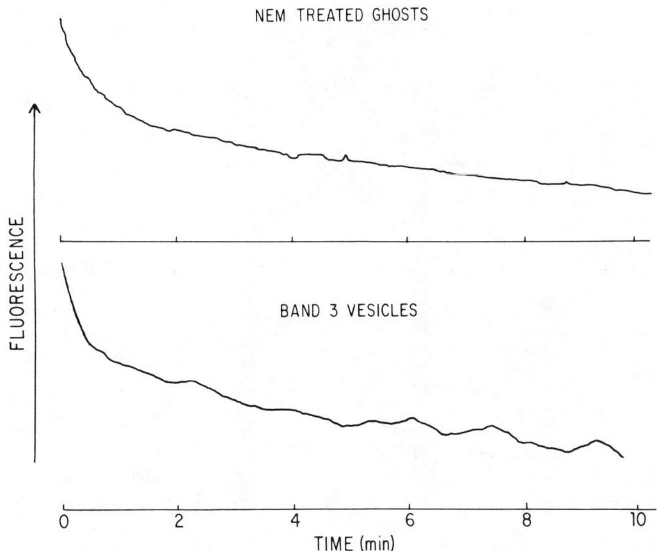

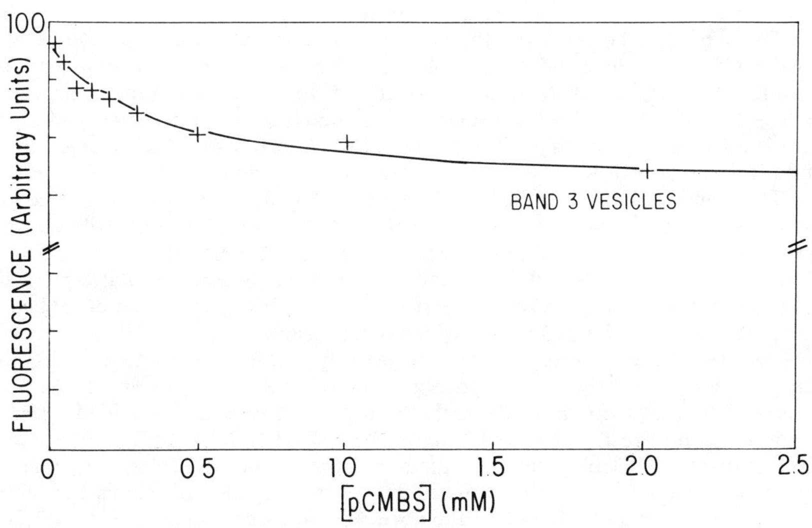

FIGURE 4. Effect of pCMBS on tryptophan fluorescence. (Top) Tryptophan fluorescence of NEM-treated red cell ghosts and band 3 vesicles (excitation 290 nm, emission 330 nm) induced by pCMBS. The initial data point for ghosts and vesicles has been set arbitrarily at the same fluorescence intensity and the amplification is the same; the initial fast drop in intensity is not shown. The pCMBS concentrations were: ghosts (10 mM) and vesicles (4 mM). As the bottom figure shows, the effect saturates at [pCMBS] ≈ 2 mM in vesicles. (Bottom) Dependence of equilibrium fluorescence intensity (measured at 30 min) of reconstituted band 3 vesicles on pCMBS concentration.

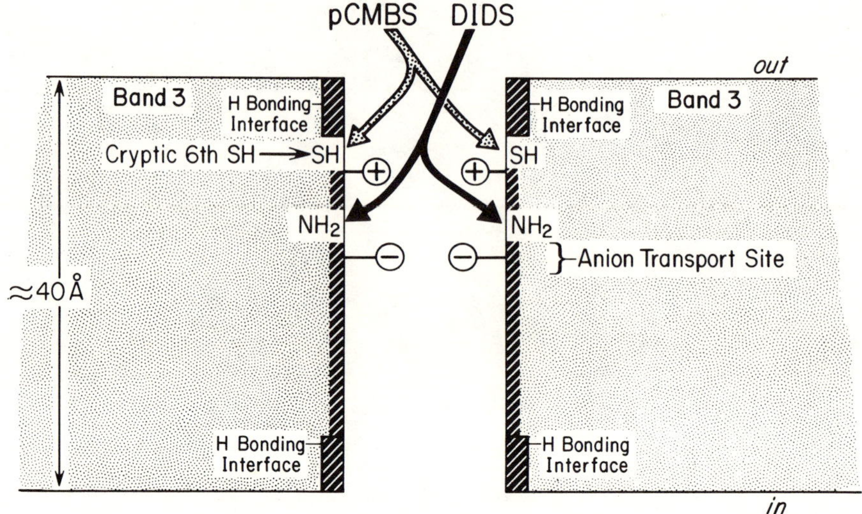

FIGURE 5. Schematic drawing of aqueous pore, located between the two monomers of a band 3 dimer, showing the relative position of the sites for DIDS and pCMBS binding and the positive and negative charge barriers. The figure also shows the hydrogen-bonding interfaces at the entrance and exit of the pore, which are linked by the polar groups that line the pore.

The similarity between the time course and concentration dependence of the pCMBS effects on the band 3 tryptophan residues and the pCMBS-induced inhibition of water flux, which takes some 10–20 min to reach maximum effectiveness[17] and has a K_I of 0.1 to 0.2 mM, is suggestive though it does not necessarily imply that there is a causal relationship between the tryptophan fluorescence changes and the inhibition of water flux.

The cryptic SH group is the key to our aqueous pore model for membrane transport of anions, cations, and small nonelectrolytes. In the sections that follow, we will show first that pCMBS binding to the SH group inhibits stilbene anion transport inhibitor binding. Next we will show that the cryptic SH group modulates passive cation transport. FIGURE 5 is a schematic drawing of an aqueous pore, located between the two halves of the band 3 dimer, which incorporates these sites. The channel contains two sequential charge barriers: a positively charged region to control cation transport and a negatively charged region to control anion transport. An unobstructed 8–9 Å aqueous pore through the red cell membrane should discriminate among small nonelectrolytes according to their size.[7] In addition to this steric discrimination, nonelectrolyte and hydrated ion fluxes can be modulated by hydrogen-bonding interactions with the walls of the pore. The two hydrogen-bonding interfaces at the entrance and exit of the pore serve as transition areas for interchange between hydration shells in free solution and hydrogen bonds within the channel, which form a continuum from one end of the channel to the other.

ANION TRANSPORT

The model in FIGURE 5 presupposes that anions and water share a common pathway in the channel between the two halves of the band 3 dimer. The anion trans-

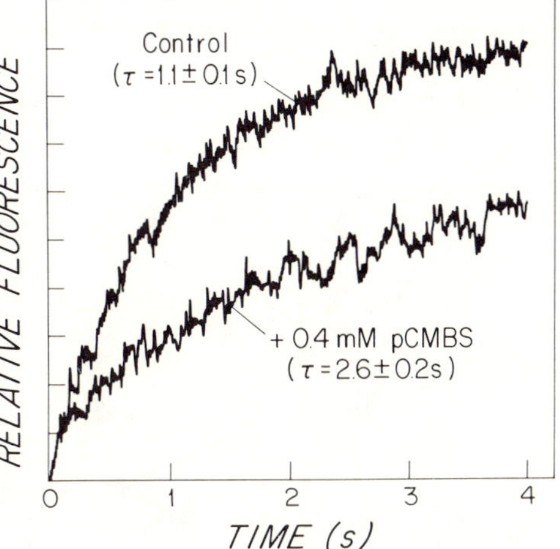

FIGURE 6. Typical experiment showing the effect of 0.4 mM pCMBS on the time course of DBDS fluorescence (emission > 410 nm; excitation 356 nm). pCMBS causes the amplitude of the signal to decrease and the time constant to lengthen.

port inhibition site in the model has been placed some 5–15 Å below the extracellular surface, which agrees with Rao *et al.*'s[65] finding that the binding site is located in a protein cleft some distance inside the membrane. The stilbene anion transport inhibitors are so large that a single molecule can just squeeze through the pore, yet DBDS does not inhibit water transport. Dix *et al.*[66] and Verkman *et al.*[67] have shown that the initial step of binding of the fluorescent stilbene inhibitor, DBDS, to the first band 3 monomer is followed by a conformational change that must take place before a second DBDS molecule binds to its inhibitory site on the second band 3 in the dimer. We presume that this conformational change involves a rearrangement of band 3 to internalize the DBDS and thus leaves the way free for the second DBDS to gain access to its site in the aqueous pore. The second DBDS would, in turn, be internalized by the second band 3 monomer so that the channel would still remain relatively open for water transport. Kleinfeld *et al.*[68] have presented evidence, based on energy-transfer measurements, that the conformational change involves a shift in the position of the tryptophan residues in band 3. Furthermore, Snow *et al.*[69] showed that stilbene inhibitors caused a large dose-related shift in one of the transitions in the calorimetric profile of band 3, which they interpreted as evidence of a conformational change in band 3. These arguments do not demonstrate that water and anions flow through the same aqueous channel; rather they provide a reasonable explanation as to how this may be possible.

Since there is evidence that binding of pCMBS to the cryptic sixth SH group in band 3 inhibits water transport, we looked for an effect of pCMBS on DBDS binding. Ghosts were treated with NEM to block the five NEM-reactive intracellular SH sites. The rate of DBDS binding to its site on band 3 was then measured, as previously described[67]; we found that pCMBS competes with DBDS binding (K_I = 0.15 mM ± 0.03).[63] FIGURE 6 shows the effect of pCMBS on the relative fluorescence of DBDS in a stopped-flow experiment and FIGURE 7 shows the concentration dependence of the pCMBS effect on the kinetics and equilibrium of DBDS binding. In order to confirm that pCMBS was bound to a SH group, we washed the preparation to remove unbound pCMBS and found that the effect of pCMBS could then be reversed by 5 mM cysteine or 5 mM glutathione, similar to the reversal of the pCMBS effect on water flux reported by Sha'afi and Feinstein.[20] Since DBDS is not a sulfhy-

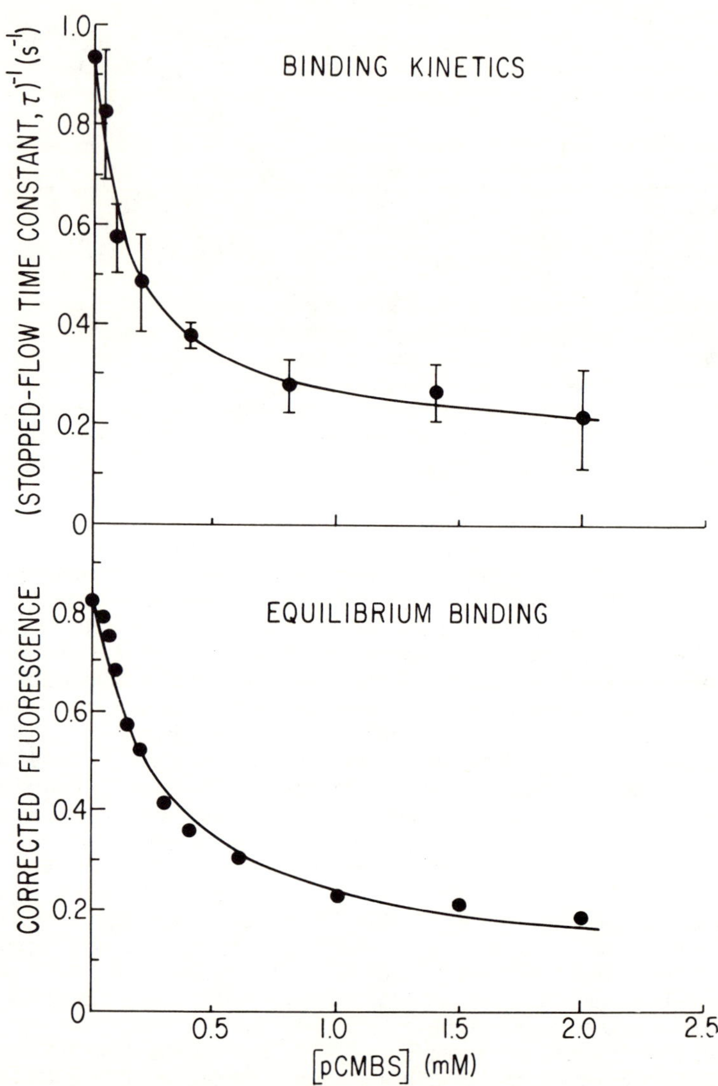

FIGURE 7. Inhibition of DBDS binding by pCMBS. (Top) Dependence of the stopped-flow time constant on the DBDS concentration in a suspension of NEM-treated red cell ghosts. Red cell ghosts were prepared by the method of Dodge *et al.*[107] and treated with 2 mM NEM for 1 hr at 25°C.[52] The stopped-flow time constant was measured by DBDS fluorescence as described by Verkman.[67] (Bottom) Inhibition of equilibrium binding of DBDS by pCMBS. Relative fluorescence was measured (emission 423 nm; excitation 360 nm) in a Perkin-Elmer Model MPF-2A spectrofluorimeter on NEM-treated ghosts, prepared as described above. The sample contained 0.04 μM band 3 (computed on the basis of 3.2 mg ghost protein/1 nM band 3 monomer) and 0.01 μM DBDS in 28.5 mM sodium citrate, pH 7.40.

dryl reagent and pCMBS binds to a SH group, we conclude that these two inhibitors do not compete simply for the same chemical group on band 3.

Many inorganic anions and a broad spectrum of organic anions use the anion-transport system (though glycolate and lactate fall into a separate class.[70]) When this diversity is examined in terms of an aqueous channel, a relatively simple mechanism for a generalized anion-transport system becomes attractive. It has been suggested by Knauf,[25] Passow *et al.*,[71,72] Macara and Cantley,[73] and Rothstein and his group[57] that the transport function of band 3 is served by moving a barrier within an aqueous pore. The barrier is considered as being operated by a flip-flop (ping-pong; *cis-trans*) gate, with the flip state facing the extracellular face, and the flop state facing the intracellular face. Anion transport takes place sequentially; when an external anion is transported the switch makes the flip-flop transition and can only be operated next by an internal anion, which resets the switch as it traverses the channel. Rothstein and his colleagues[22,25] showed that there are inside- and outside-facing conformations of the transport protein and Gunn and Fröhlich[74,75] have shown that these conformations are asymmetrical with respect to anion-binding properties. Jennings[76] has provided evidence that Cl^- influx and efflux are sequential, rather than simultaneous, processes. However, a movable barrier or even a movable charged group, as suggested by Passow *et al.*,[71] can not easily exercise selectivity among the diverse anions that enter the cell. In our model, the selection process would be exercised by the walls of the pore. Thus, there would be two major determinants of passage through the anion channel. The discrimination among anions would depend upon interactions with the walls of the aqueous channel, which would sieve out large anions and modulate the transport of those small enough to permeate by selective hydrogen-bonding sites on the walls of the pore. The second barrier would depend solely on charge of the anion, which would be sensed by band 3 and cause it to set the switch in its flip or flop position, depending on the direction from which the anion came.

The apparent independence of the transport properties of the two monomers of band 3 introduces an additional restraint in the mechanism. The size of the hydrated Cl^- ion[77] is such that it does not seem possible for an ion moving in one direction to get past one traveling in the opposite direction. This constraint means that only one ion should be in the pore at any one time. The volume of the pore is so small that a solution of 155 mM KCl is equivalent to an average density of 0.24 Cl^- ions/pore, assuming equal concentration in the pore and in the solution; on this basis 75% of the pores are always free of Cl^- ions. The ion flux of 4.9×10^{10} ions/cell, sec at 38°C given by Brahm[78] is equivalent to a flux of 0.9×10^5 ions/pore, sec on the basis of 5.5×10^5 dimers of band 3/cell. On the average, one ion is discharged from each pore every 11.2 μsec, an average transit time that includes both the time to diffuse through the pore and the time required for the flip-flop reaction. The diffusion time can be approximated from the equation: $2 D_{KCl} t$ = (channel length)^2. Using $D_{KCl} = 2.6 \times 10^{-5}$ cm^2 sec^{-1} at 38°C[79] and a 40 Å pore length, t = 3.1 nsec. This minimum estimate must be increased by a steric hindrance factor of perhaps one order of magnitude.

The gating-time restriction, which is even more stringent, arises because the steric constraints of the pore make it impossible for both monomers to transport a Cl^- ion at the same instant. It is known, however, that each monomer can independently transport ions. This observation places an upper limit on the time it takes the flip-flop to operate, the gating-time, which must in any case be shorter than the transit time. This problem is formally similar to the coincidence problem in particle counting. We shall impose the requirement that the dependence of Cl^- flux on the numbers of functioning band 3 monomers is linear to within 2%, which is consistent with the data of Lepke *et al.*[80] The coincidence equation¶ then imposes a limit on the gating-time of operation of the flip-flop of $\tau = 0.4$ μsec. Since chemical reactions commonly have time constants of 10^{-8} sec and faster, a gating-time of 0.4 μsec is not chemically unreasonable. The transit time is slower than the gating-time by almost two orders of magnitude, which allows for significant steric hindrance. These calculations indicate that the aqueous channel can meet the requirement of independent monomer action and still accommodate the observed Cl^- flux.

Since the flip-flop conformational change depends upon electrostatic interactions, common to all anions of the same charge, and not upon specific steric and hydrogen-bonding characteristics of the transported ion, we would expect that the Arrhenius activation energy of anion exchange would be essentially independent of the chemical identity of the transported anion. Dalmark and Wieth[81] have shown that the activation energies of the transport of Cl, Br, CNS, and I are similar, lying between 29 and 37 kcal/mol, and Aubert and Motais[82] have shown that similar values are characteristic of organic anion permeation. It is difficult to account for this uniformity in a diverse set of specific interactions involving anions of widely different chemical composition, but easy to consider it a consequence of a flip-flop gate whose mechanism is governed by the charge of the transported ion. The further steric and polarity requirements discussed by Aubert and Motais[82] would be imposed by the hydrogen bonding and other characteristics of the wall of the pore. Thus, the transport mechanism would incorporate two separate properties of the pore, one to provide the specific molecular restraints characteristic of the transported ion and another to control the gate.

Cation Transport

pCMBS causes a massive red cell cation leak.[84,85] Grinstein and Rothstein[86] showed that this pCMBS effect was preserved in inside-out vesicles that had been stripped of bands 1, 2, and 5 and considerably depleted in band 6. Indeed, the pCMBS effect on cation fluxes could still be demonstrated in these vesicles when they had been treated with chymotrypsin and trypsin so that the major remaining sequences of band 3 in these vesicles were the 17 kd fragment and the 8–9 kd membrane loop.[59] We have interpreted these experiments in terms of the model shown in Figure 5, as consistent with our view that the cation leak is modulated by a pCMBS site in band 3 outside of the stilbene inhibition site, but in close proximity to it. This position of the pCMBS site provides an alternative explanation for Knauf and Rothstein's[51] observation that stilbene inhibitors did not affect the pCMBS-induced cation leak. They interpreted their finding as evidence for an alternative pCMBS path-

¶We assume that each monomer participates in the anion-transport process and that both flip-flop gates are located in the same channel, but only a single gating-step can take place at one time. Each monomer can process 4.5×10^4 Cl^- ions sec^{-1}. We may ask: what is the transition time for the rate-limiting step if that step is able to process essentially all the anions passing through the channel? The coincidence equation is $N_C = N_1 N_2 \tau$ in which N_C is the number of coincidences per sec, N_i is the number of particles passing through the ith channel per sec, and τ is the coincidence time. Suppose that 2% of the anions don't get through because the rate-limiting step is "busy." Then $N_1 = N_2 = 4.5 \times 10^4$ sec^{-1} and the number of "coincidences" is $N_C = 9 \times 10^2$ sec^{-1} and therefore

$$\tau = \frac{N_C}{N^2} = 0.4\ \mu\text{sec}.$$

way into the cytoplasm that was not affected by stilbene inhibitors. Our model explains their observation by placing the cryptic sixth SH group outside of the stilbene inhibitor site.

In order to study the cation-leak process further, we have measured pCMBS-induced cation leakage in red cells in which the five NEM-reactive SH groups were blocked. Under these conditions pCMBS still induces a substantial cation leak.[87] Since this is not an inhibitory process, there is no K_I, but the pCMBS effect is substantial for pCMBS concentrations in the range of 0.05 to 0.2 mM consistent with the values of K_I of 0.1–0.2 mM obtained for the effects of pCMBS on water flux and stilbene inhibitor binding. As a further test of the specificity of this SH group, we have examined three other sulfhydryl reagents, whose effects on water transport have been determined by Sha'afi and Feinstein;[20] *p*-chloromercuribenzoate (pCMB) and *p*-aminophenylmercuric acetate (pAPMA), which inhibit water transport by about 80%, and iodoacetamide (IAM), a non-mercurial sulfhydryl reagent that does not affect water transport. As FIGURE 8 (left) shows, both pCMB and pAPMA induce large cation leaks in NEM-treated cells, whereas IAM is without effect. As FIGURE 8 (right) shows, the mercurials that inhibited water transport and induced cation leakage also produced effects on DBDS binding in NEM-treated red cells, whereas IAM which had no effect on water transport had no effect on the other systems. The experiments with these reagents, therefore, provide strong support for the view that the same cryptic SH group is responsible for the effects we have observed in all three systems.

These experiments show that an NEM-insensitive SH group is responsible for the cation leak, but they do not localize the SH group to band 3, since there are SH groups on other integral membrane proteins, such as the (Na,K)-ATPase and band 4.5, which might be responsible for the effect. We therefore carried out experiments on separated band 3, incorporated into phosphatidylcholine vesicles. Band 3 was isolated by the method of Lukacovic *et al.*,[64] which provides 96% pure band 3 containing no band 4.5. The results of one cation-leak experiment, typical of four, are given in FIGURE 9, which shows that pCMBS induces a cation leak in band 3 vesicles. Control experiments showed that this induced leak could not be attributed to (Na,K)-ATPase. In other experiments in the same series, treatment with 2 mM NEM for 10 min at 23°C did not induce a cation leak. Also 5 mM mercaptoethanol reversed most of the pCMBS-induced leak, showing that it is a SH group with which the pCMBS reacts. Since NEM alone has virtually no effect, none of the intracellular SH groups on band 3 can be responsible for the pCMBS-induced cation leak, which may therefore be attributed to the cryptic, sixth SH group in the 17 kd membrane fragment.

HYDROGEN-BONDING INTERACTIONS WITH THE PORE WALL: ANIONS

Clementi *et al.*[77] report the radius of the first hydration shell of Cl^- to be 3.9 ± 0.4 Å, which is comparable with the pore radius, so that interactions between the hydration shell and hydrogen-bonding sites on the channel are probably required and may well account for selectivity among inorganic anions. The principles are similar to those discussed by Hille[88] for the Na channels in nerve, but the red cell aqueous pore diameter is about twice as large as that of the nerve Na channel, consonant with the difference between the stringent selectivity exercised by the Na channel and the broad spectrum of inorganic anions that use the red cell anion transport system. Alternatively, the Cl^- exchange transport system could move the Cl^- ion by transporting it in the lipid phase for some or all of the 40 Å distance across the lipid bilayer.

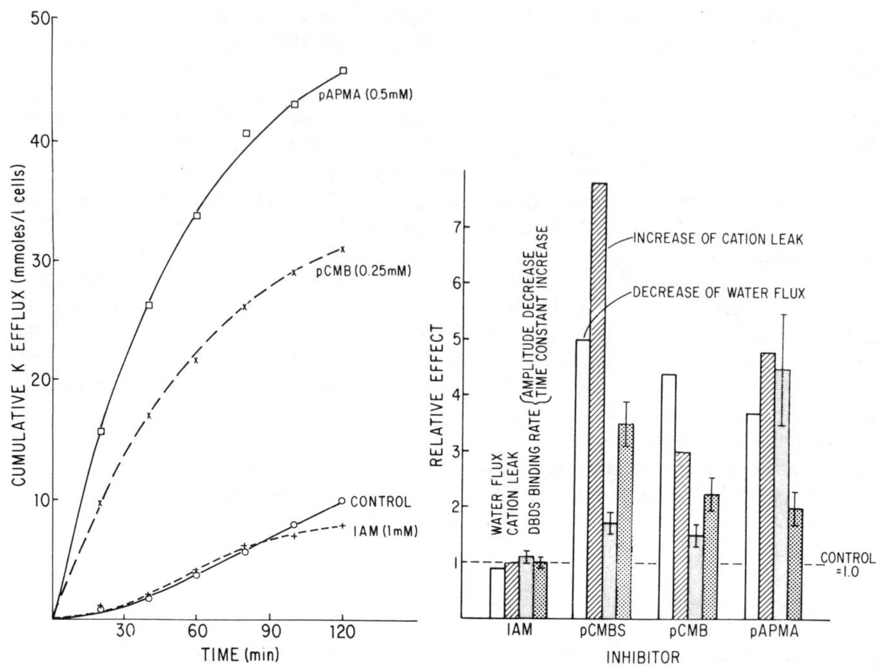

FIGURE 8. Cation leak induced in red cells by sulfhydryl reagents. (Left) Uptake of ^{86}Rb in NEM-treated red cells. K efflux was measured in red cells from freshly drawn human blood treated with NEM (12 mM, 1 hr, 37°C) in order to block all NEM-reactive sulfhydryls. Fluxes were measured with ^{86}Rb, as described by Poznansky and Solomon.[108] The cumulative K efflux was computed as the sum of the influx measured with ^{86}Rb and the net K efflux measured chemically. No correction was made for the difference in influx resulting from the changing extracellular K concentrations at the several inhibitor concentrations. The concentration of the inhibitor in the cation-leak experiments was limited by hemolysis of the NEM-treated cells, so that the concentrations of the reagents that accelerated the leak (pCMBS, 0.2 mM; pCMB 0.25 mM; pAPMA, 0.5 mM) were less than that of IAM, which had no effect at 1 mM. The curves were drawn by eye to connect the points. (Right) Comparison of effects of sulfhydryl reagents on different transport processes. Four reagents were chosen from those whose effects (at 1 mM) on water flux had been measured by Sha'afi and Feinstein,[20] three that inhibited water flux (pCMBS, pCMB and pAPMA) and one (IAM) that had no effect. The bar graph has been drawn so that effect of the sulfhydryl reagent, whether enhancement or inhibition, is shown by an increase in the length of the bar. The control value of 1.0 represents no effect. DBDS binding was studied at 1 mM and the effect was determined from the amplitude and the time course of the DBDS reaction with its binding site as described by Verkman and Dix.[109] The water flux data are taken from Sha'afi and Feinstein (Table 2 in Reference 20).

This seems a relatively unlikely mechanism since it would be difficult both to meet the transit-time restraints and to provide the 37 kcal/mol of potential energy[89] required to insert the Cl^- ion into the lipid bilayer.

There is great diversity among the organic anions that use the ion-transport system, which includes a molecule as small as formic acid with a half-time of permeation of 0.3 min and one as large as *p*-aminohippuric acid with a half-time of 90 min. *p*-Aminohippuric acid is the largest molecule that Aubert and Motais[82] found to use the anion channel. We have made a CPK model of it and found that it is small

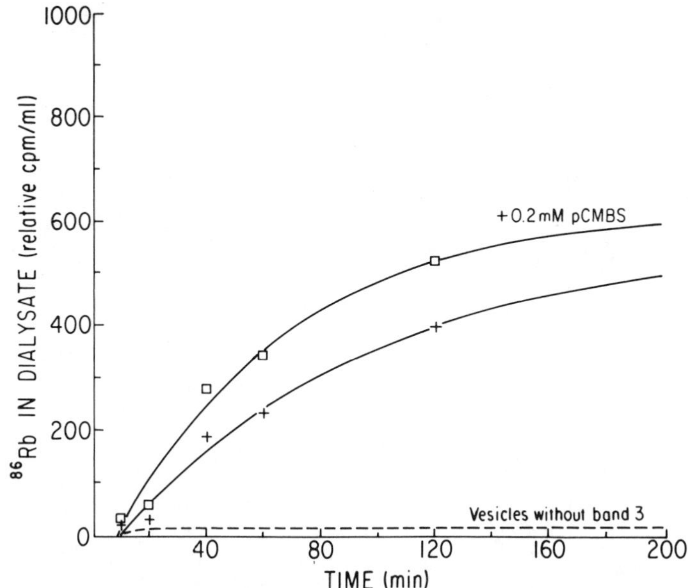

FIGURE 9. Effect of pCMBS on cation leak in reconstituted band 3 vesicles. Purified band 3 was reconstituted into phosphatidylcholine vesicles, which were placed in dialysis bags. ^{86}Rb efflux was measured at 23°C as described for sulfate efflux by Lukacovic *et al.*[64] The curves are drawn by eye (control, +; pCMBS treated, □). The data for uptake in vesicles without band 3 were taken from other experiments in the same series. The figure is typical of four experiments with pCMBS, in two of which NEM and mercaptoethanol were added.

enough to go through an 8–9 Å diameter pore. Aubert and Motais[82] have suggested that there are specific structural requirements governing the transport of this diverse group of solutes, which have a common requirement for "at least a three point attachment involving three oxygen atoms in the substrate." As previously pointed out, our model would satisfy these requirements by hydrogen-bonding sites on the wall of the pore.

HYDROGEN-BONDING INTERACTIONS WITH THE PORE WALL: NONELECTROLYTES

Urea transport into the red cell is a saturable process with an apparent K_D of 0.1–0.3 M, which can be inhibited by thiourea with a K_I of about 10 mM (FIGURE 10, top).[47,90–92] These observations have been interpreted as support for Macey *et al.*'s[17] suggestion that urea enters the red cell by facilitated diffusion.[93] However, as Hille[88] has pointed out, saturation and competitive inhibition are also characteristic of a saturable pore and can not be used to discriminate between transport through a pore and a facilitated-diffusion model. According to our model, the hydrogen-bonding regions shown in FIGURE 5 facilitate the interchange of solute hydration shells with hydrogen bonds within the pore, which form a continuous pathway through the membrane. We ascribe the saturation and apparent competitive inhibition of urea transport to a weak binding site for urea in the hydrogen-bonding interface regions. Urea has an anomalously small partial molar volume, presumably reflecting its ca-

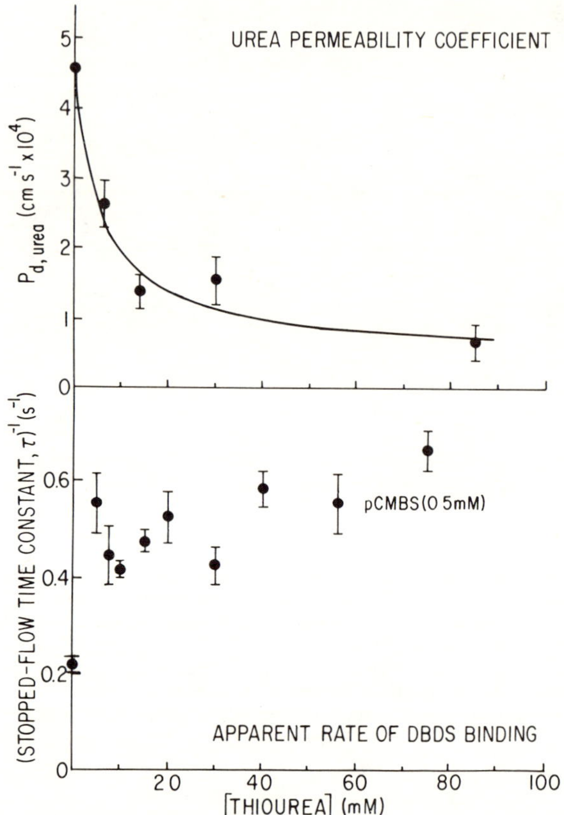

FIGURE 10. Effect of thiourea on urea flux into red cells and on pCMBS competition with DBDS binding. (Top) Urea permeability was measured by the method of Sha'afi *et al.*[110] at a urea concentration of 0.7 M. Data are presented for one experiment, typical of three. The points were fitted to a single-site model by least squares and give $K_{1/2} = 5 \pm 1$ mM. (Bottom) The effect of thiourea on the pCMBS inhibition of the rate of DBDS binding.

pacity to form six hydrogen bonds and therefore interact strongly with water. Passage through the pore would require replacement of this hydration shell by hydrogen bonds in the pore. Thiourea could inhibit urea transport by competing at the urea hydration shell interchange site; since the ether/water partition coefficient of thiourea is more than an order of magnitude larger than that of urea,[94] thiourea binding could well be enhanced by a neighboring lipid region.

We reasoned that thiourea binding to the wall of the aqueous channel might produce effects on other features of the transport system and found, as shown in FIGURE 10, that thiourea reverses the pCMBS inhibition of DBDS binding. Control experiments show that thiourea has no direct effect on DBDS fluorescence. The figure indicates that K_I for reversal is in the range of 0–5 mM, below the value of K_I = 10 mM for urea transport inhibition given by Mayrand and Levitt.[90] This observation indicates that thiourea, which specifically inhibits urea and amide diffusion, is a more general reagent than had hitherto been thought, since it interferes with the pCMBS-DBDS binding interaction.

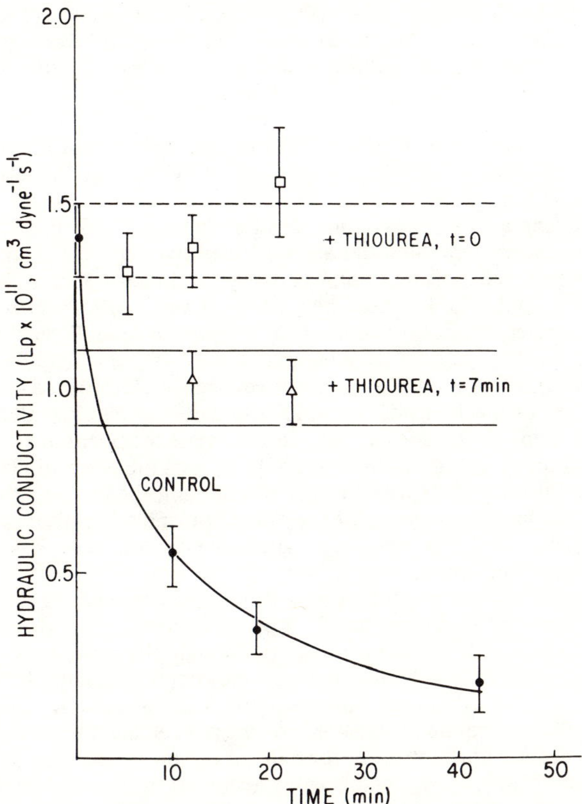

FIGURE 11. Time course of pCMBS inhibition of osmotic water flux. The concentration of pCMBS was 1 mM and that of thiourea, 50 mM. pCMBS was added at the beginning of the experiment and thiourea was added either at the beginning or 7 min later.

The interaction between thiourea and pCMBS reflected in the apparent rate of DBDS binding could be explained on the basis of a thiourea-induced distortion of the pCMBS binding site. This consideration led us to look for other relations between thiourea and pCMBS. We found that 50 mM thiourea turns off the pCMBS inhibition of water transport (FIGURE 11). If pCMBS and thiourea are added together at the beginning of the experiment, no pCMBS inhibition develops.** If the

** One possible explanation for the thiourea-pCMBS effects would be a chemical reaction between thiourea and the Hg in pCMBS. We tested this possibility by using the yellow color developed by DTNB (5,5′-dithio-bis-(2-nitrobenzoic acid)) upon reaction with cysteine. pCMBS inhibits the color development, if added in equimolar amounts with the cysteine. If thiourea formed a complex with pCMBS tighter than that formed between cysteine and pCMBS, addition of thiourea to the reaction mixture would interfere with the inhibition of color development that pCMBS normally causes. This was not found to be the case since no interference was observed when excess thiourea was preincubated with the organic mercurial for 30 minutes. In similar experiments using DTNB, it was also shown that thiourea does not react with cysteine. This result indicates that the thiourea effect on band 3 does not proceed by competition of thiourea for the cryptic sixth SH group with which pCMBS reacts. Two further lines of evidence thiourea is added later, it stops the pCMBS inhibition dead in its tracks, not only at 7 min (FIGURE 11), but also at other times during the 30-min time course required for full development of the pCMBS inhibition of water transport. Although we do not yet know the exact K_I for this process, preliminary experiments suggest a K_I in the range of 1–3 mM. The observation that K_I for the thiourea release of the pCMBS inhibition of water transport is different from the K_I for the inhibition of urea flux means that the thiourea-pCMBS interaction cannot be explained in terms of a single binding site producing a single effect.

There are unexplained observations about the action of pCMBS and other mercurial sulfhydryl reagents on the water and nonelectrolyte systems. Though the effect of pCMBS on water transport has a time course extending for 30 min, both Hg^{2+} and pCMB, which are lipid soluble, exert their action in 5 min or less. The pCMBS inhibition of urea permeability also is complete in 5 min or less. Typically, reversal of pCMBS inhibition by cysteine or mercaptoethanol is very fast for water flux, whereas it is very slow for urea. We have not yet been able to form a coherent model of the mechanism, or mechanisms, by which these mercurials exert their effect. The absence of a coherent model does not, however, vitiate the phenomenological observations that demonstrate that thiourea not only inhibits the permeability of urea and other amides but also interacts with the pCMBS binding site in a way that affects the transport of water and the binding of the anion transport inhibitor, DBDS. These actions are consistent with our view that a common element, band 3, is involved with all of the processes concerned.

The differential effects of phloretin on water and urea permeability can also be understood in terms of hydrogen-bonding properties. Macey and Farmer showed that 0.5 mM phloretin inhibits urea transport, though phloretin has virtually no effect on water transport.[16,17,95,96] Jennings and Solomon[97] measured the binding of phloretin to red cell membrane components at pH 6.0 and found K_D to be 1.5 μM for the red cell membrane proteins and 54 μM for the red cell membrane lipids. We have measured the phloretin concentration for half-inhibition of urea transport at pH 7.4 and found it to be 70 μM, in reasonable agreement with the value of 90 μM for $K_{D,\text{lipid}}$ obtained by converting Jennings and Solomon's K_D to pH 7.4. These findings are consistent with the view that phloretin exercises its effect indirectly through the lipid binding region and not directly through binding to a membrane protein. This conclusion is also consistent with Snow *et al.*'s[69] observation that 0.21 mM phloretin affects the heat-capacity profile of the red cell membrane in a region that these authors ascribe to band 3/phospholipid interaction. Evidence that perturbations in the lipid can affect band 3 conformation is afforded by Forman *et al.*'s[98] demonstration that DBDS binding kinetics are dependent upon the membrane solubility[99] of alcoholic anesthetics.

Owen and Solomon[95] and Owen *et al.*[96] pointed out that phloretin inhibits the permeability of many hydrophilic nonelectrolytes and that the degree of inhibition is inversely proportional to the ether-water partition coefficient, which is an index of hydrogen-bonding capacity. This apparent dependence upon hydrogen bonding is consistent with a phloretin-induced perturbation of the hydrogen-bonding region.

---

support these conclusions. Addition of 10 mM thiourea to the ghost suspension, which was analyzed by polyacrylamide gel electrophoresis as described in FIGURE 3, had no measurable effect on the [^{203}Hg]pCMBS distribution. The ^{13}C NMR spectrum of 100 mM thiourea was examined in a 360 MHz NMR spectrometer (Fossel, private communication). Addition of 20 mM pCMBS had no effect on the ^{13}C resonances.

Molecular Model of the Aqueous Pore

A molecular model of the aqueous pore can be constructed from the elements of band 3 most closely concerned with transport, the 15 kd membrane-transport fragment and the 8 kd membrane loop that Ramjeesingh *et al.*[54,59] have analyzed. Guidotti[100] has made two important observations that have been instrumental in constructing our model; first that those segments of membrane proteins that run through the membrane could be expected to have a high α-helix content,[101] and second, that there are great similarities in the structure of the intrinsic proteins that span membranes. There are 137 residues in the 15 kd membrane transport fragment of band 3 and 84 residues in the membrane loop. Since the residues form a continuous sequence from the N terminus on the 15 kd segment to the C terminus of the 8 kd segment and since the N and C termini are on opposite sides of the membrane, this sequence must cross the membrane an odd number of times.

Our structure has been modeled on that of bacteriorhodopsin given by Engelman *et al.*,[102] which comprises a compact assembly of seven α-helical regions lacing through the membrane. Six of the seven α-helical chains in bacteriorhodopsin average 26 residues in length, including one extra residue at each terminal of the α-helical segment; each segment is about 10 Å in diameter. The connections between the segments are of varying length; the shortest requires only the single residue at the end of each helix that is included in the 26-residue figure; the longest includes 9 additional residues. There are a total of 247 residues in the seven α-helices, or 35 residues per α-helical leg. If the 221 residues in the membrane-transport sequence plus the membrane loop were composed similarly of α-helical legs, five α-helices would average 44 residues per leg and seven would average 32 residues per leg.

In constructing our molecular model for the aqueous pore, we have been guided by the following set of basic principles. (1) The aqueous pore is located between two monomers in a dimeric structure because most known examples of aqueous channels are formed between oligomers of membrane proteins (Wiley, personal communication). (2) The membrane transport assembly is presumed to consist of a number of α-helical legs, similar to bacteriorhodopsin, with an odd number of crossings in each monomer. (3) The hydrophilic residues of the α-helical assembly are in contact with the aqueous channel and the hydrophobic residues stabilize the assembly by contacts between the α-helical legs and the membrane lipids. (4) The model must conform to the restraints that the cryptic sixth SH is outside of the DIDS site and that the two SH groups on the membrane loop are so close that they can cross-link readily.[50]

Since band 3 is a non-covalent dimer, we explored the various ways in which an odd number of α-helical runs in each monomer might be combined to form a pore of about 9 Å diameter. A particularly attractive structure is the symmetrical hexagonal array (shown at the bottom right of FIGURE 12) that comprises an inner ring of α-helices 1, 2, and 4, taken twice, that surround the space reserved for the central α-helix. By removing the central α-helix and considering the vacated space as the pore, a symmetrical structure could be constructed in which a 10 Å pore was formed between the two halves of the dimer. The array would be stabilized by the hydrophobic and hydrophilic contacts described above.

We have numbered the residues consecutively from the N terminal, based on an average of 108 daltons per residue, as shown in FIGURE 12. Ramjeesingh *et al.*[54] have located the H_2DIDS binding site of band 3 in a 2 kd segment between residues 54 and 73 as shown in the kilodalton map at the top of FIGURE 12. As previously stated, Rao *et al.*,[65] who have shown that the DIDS site is located some 34 to 42 Å away from the SH sites on the 43 kd cytoplasmic N-terminal fragment, suggest that the DIDS site is located in a cleft on the extracellular surface of the membrane. Rothstein and Ram-

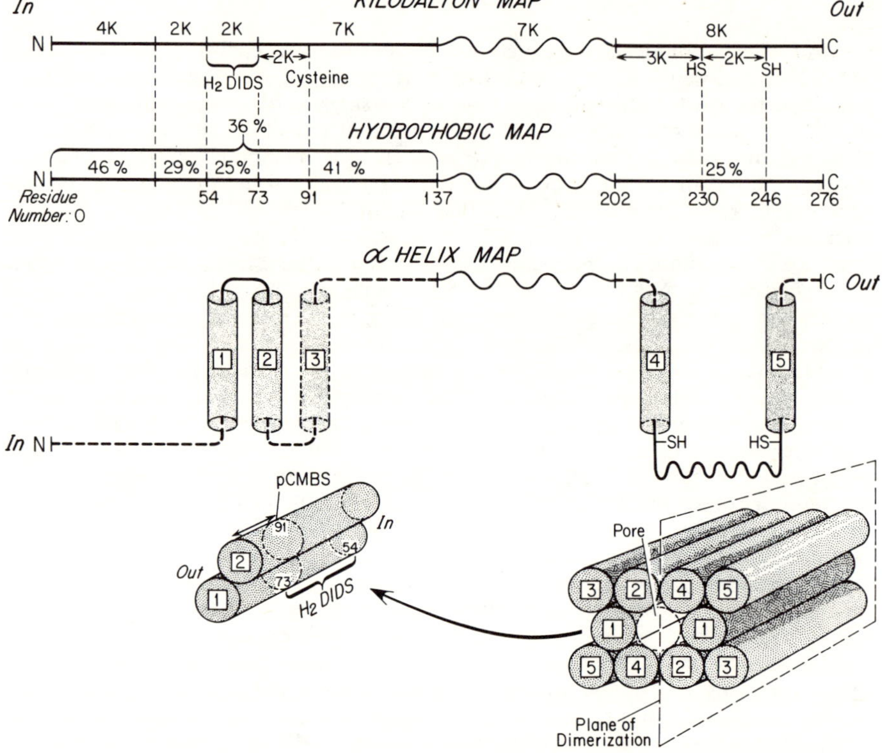

FIGURE 12. Suggested α-helical model of aqueous channel in red cell membrane.

jeesingh[57] have located the cryptic sixth SH group 2 kd further along the transport fragment, which corresponds to residue 91 in FIGURE 12 by our calculation.

We have placed these sites on adjacent α-helices, as shown in the α-helix map in FIGURE 12, for the following reasons. The minimum distance between the cryptic sixth SH group and the H_2DIDS site is 18 residues, which corresponds to about 28 Å if the protein is in an α-helical conformation. Assuming that the site of the anion-transport inhibition is on helix 1 within a few angstroms of the H_2DIDS reaction site, the Rao *et al.* experiments place the site no more than about 10 Å below the extracellular face. Thus it is not possible for the cryptic sixth SH to be on the same helical leg; we have placed it on the adjacent helix 2, which requires that the junction between helix 1 and helix 2 be at the outer face of the membrane. The actual depth of the cryptic sixth SH within the pore is determined by the length of the connector stretch between helices 1 and 2.

Rothstein and Ramjeesingh[57] have shown that the first SH group in the 8 kd membrane loop (computed to be residue 230) is 3 kd from the external N-terminus of the loop; the second SH is 1.7 kd further along (residue 246) and the C-terminus of the loop is extracellular. Three kilodaltons are equivalent to 28 residues, just long enough for a α-helix to reach from the extracellular N-terminus of the membrane loop to an SH placed on the intracellular face. The 1.7 kd stretch between the two SH groups (rounded to 2 kd in FIGURE 12) may be presumed to be in the cytoplasm. The remaining 3.3 kd in the membrane loop permits an α-helical leg to run from the second SH placed on the intracellular face to the extracellular C-terminus of the membrane loop. This arrangement places the two SH groups close together on the intracellular face and is well adapted to produce the spontaneous cross-linking that Rao and Reithmeier[50] have observed in red cell ghosts. Since the C-terminus of helix 2 (considered as a fragment) is inside the membrane and the N-terminus of helix 4 is outside the membrane, another crossing is necessary. We assume this crossing to go via α-helix 3, which must lie between residues 91 and 137 in the membrane-transport sequence. α-Helix 3 has no known transport function and we have placed it in the lipid and denoted it by a dashed line rather than a full line in FIGURE 12.

Our model fits moderately well with the restraints offered by the polarity[103] of the several sequences as shown in the hydrophobic map of FIGURE 12. The most polar of the residues in the 15 kd membrane-transport sequence are in the two central 2 kd segments. They are presumed to constitute a significant portion of α-helices 1 and 2, which are in contact with the aqueous channel in the pore. Residue 91 in the 41% hydrophobic 7 kd fragment is also in the aqueous channel but is counterbalanced by α-helix 3, which runs through the lipid. The membrane loop is only 25% hydrophobic. This can be accounted for, in part, by the cytoplasmic location of the stretch from residues 230 to 246 and the fact that helix 4 is also considered to be in contact with the aqueous pore. If the hexagonal symmetry is to be preserved, helix 5 must be in the lipid domain.

The picture that emerges incorporates band 3 as the centerpiece of a macromolecular array of proteins responsible for governing the passive transport of water, nonelectrolytes, anions, and cations into and out of the red cell. The residues on the α-helices in contact with the aqueous pore would be specified genetically to form the hydrogen bonds that govern the selectivity found in organic anion and nonelectrolyte transport. Although a significant number of questions remain to be resolved, the experiments reported here support many of the interlocking features required by the model, which would provide a delicately organized and easily controlled system of governance of the traffic across the membrane.

ACKNOWLEDGMENTS

The authors would like to express their gratitude to V. Rhoden for expert technical assistance and to Mr. Bernard Corrow and Dr. Alfred Pandiscio for their essential help in the design and construction of the stopped-flow and temperature-jump equipment. We are particularly grateful to Prof. A. Rothstein who not only shared his unpublished results with us, but who also carried out so much of the original work on the effect of sulfhydryl reagents on which our findings were based. The original impulse to identify the sulfhydryl group responsible for the inhibition of water transport grew out of a conversation with Prof. G. Guidotti whose ideas about the role of the α-helix in membrane transport were also of great importance to us. We should also like to thank Dr. Anjana Rao since the unpublished data in her thesis on pCMBS membrane sulfhydryl reactions were basic to our studies.

Dr. Chasan wishes to acknowledge a sabbatical leave granted by Boston University and Dr. Lukacovic, a research fellowship from the American Heart Association, Western Massachusetts Division, number 13-401-812.

REFERENCES

1. PAGANELLI, C. V. & A. K. SOLOMON. 1957. The rate of exchange of tritiated water across the human red cell membrane. J. Gen. Physiol. **41**: 259–277.

2. Sidel, V. W. & A. K. Solomon. 1957. Entrance of water into human red cells under an osmotic pressure gradient. J. Gen. Physiol. **41**: 243–257.
3. Goldstein, D. A. & A. K. Solomon. 1960. Determination of equivalent pore radius for human red cells by osmotic pressure measurement. J. Gen. Physiol. **44**: 1–17.
4. Pappenheimer, J. R., E. M. Renkin & L. M. Borrero. 1951. Filtration, diffusion and molecular sieving through peripheral capillary membranes. Am. J. Physiol. **167**: 13–46.
5. Renkin, E. M. 1954. Filtration, diffusion, and molecular sieving through porous cellulose membranes. J. Gen. Physiol. **38**: 225–243.
6. Barton, T. C. & D. A. J. Brown. 1964. Water permeability of the fetal erythrocyte. J. Gen. Physiol. **47**: 839–849.
7. Solomon, A. K. 1968. Characterization of biological membranes by eqivalent pores. J. Gen. Physiol. **51**: 335s–364s.
8. Vieira, F. L., R. I. Sha'afi & A. K. Solomon. 1970. The state of water in human and dog red cell membranes. J. Gen. Physiol. **55**: 451–466.
9. Wang, J. H. 1951. Self-diffusion and structure of liquid water. J. Am. Chem. Soc. **73**: 510–513.
10. Solomon, A. K. 1972. Properties of water in red cell and synthetic membranes. *In* Passive Permeability of Cell Membranes, Biomembranes Vol. 3. F. Kreuzer & J. F. G. Slegers, Ed.: 299–330. Plenum Press. New York.
11. Mueller, P., D. O. Rudin, H. T. Tien & W. C. Wescott. 1962. Reconstitution of cell membrane structure in vitro and its transformation into an excitable system. Nature **194**: 979–980.
12. Cass, A. & A. Finkelstein. 1967. Water permeability of thin lipid membranes. J. Gen. Physiol. **50**: 1765–1784.
13. Holz, R. & A. Finkelstein. 1970. The water and nonelectrolyte permeability induced in thin lipid membranes by the polyene antibiotics nystatin and amphotericin B. J. Gen. Physiol. **56**: 125–145.
14. Solomon, A. K. & C. M. Gary-Bobo. 1972. Aqueous pores in lipid bilayers and red cell membranes. Biochim. Biophys. Acta **255**: 1019–1021.
15. DeKruijff, B. & R. A. Demel. 1974. Polyene antibiotic sterol interactions in membranes of *Acholeplasma laidlawwii* cells and lecithin liposomes. Biochim. Biophys. Acta **339**: 57–70.
16. Macey, R. I. & R. E. L. Farmer. 1970. Inhibition of water and solute permeability in human red cells. Biochim. Biophys. Acta **211**: 104–106.
17. Macey, R. I., D. M. Karan & R. E. L. Farmer. 1972. Properties of water channels in human red cells. *In* Passive Permeability of Cell Membranes, Biomembranes, Vol. 3. F. Kreuzer & J. F. G. Slegers, Eds.: 331–340. Plenum Press. New York.
18. Brown, P. A., M. B. Feinstein & R. I. Sha'afi. 1975. Membrane proteins related to water transport in human erythrocytes. Nature **254**: 523–525.
19. Naccache, P. & R. I. Sha'afi. 1974. Effect of pCMBS on water transfer across biological membranes. J. Cell Physiol. **84**: 449–456.
20. Sha'afi, R. I. & M. B. Feinstein. 1977. Membrane water channels and SH-groups. *In* Advances in Experimental Medicine and Biology. M. W. Miller, A. E. Shamoo & J. S. Brand, Eds. **84**: 67–80. Plenum Press. New York.
21. Fairbanks, G., T. L. Steck & D. F. H. Wallach. 1971. Electrophoretic analysis of the major polypeptides of the human erythrocyte membrane. Biochemistry **10**: 2606–2617.
22. Cabantchik, Z. I., P. A. Knauf & A. Rothstein. 1978. The anion transport system of the red blood cell. Biochim. Biophys. Acta **515**: 239–302.
23. Pinto da Silva, P. 1973. Membrane intercalated particles in human erythrocyte ghosts: Sites of preferred passage of water molecules at low temperature. Proc. Natl. Acad. Sci. USA **70**: 1339–1343.
24. Pinto da Silva, P. & G. L. Nicolson. 1974. Freeze-etch localization of concanavalin A receptors to the membrane intercalated particles of human erythrocyte ghost membranes. Biochim. Biophys. Acta **363**: 311–319.
25. Knauf, P. A. 1979. Erythrocyte anion exchange and the band 3 protein: Transport kinetics and molecular structure. *In* Current Topics in Membranes and Transport. F. Bronner & A. Kleinzeller, Eds. **12**: 249–363. Academic Press. New York.

26. GUIDOTTI, G. 1972. The composition of biological membranes. Arch. Intern. Med. **129**: 194-201.
27. RAO, A. 1979. Disposition of the band 3 polypeptide in the human erythrocyte membrane. J. Biol. Chem. **254**: 3503-3511.
28. CONLON, T. & R. OUTHRED. 1978. The temperature dependence of erythrocyte water diffusion permeability. Biochim. Biophys. Acta **511**: 408-418.
29. FABRY, M. E. & M. EISENSTADT. 1975. Water exchange between red cells and plasma. Biophys. J. **15**: 1101-1110.
30. MORARIU, V. V., V. I. POP, O. POPESCU & G. BENGA. 1981. Effects of temperature and pH on the water exchange through erythrocyte membranes: nuclear magnetic resonance studies. J. Membr. Biol. **62**: 1-5.
31. SHPORER, M. & M. M. CIVAN. 1975. NMR study of ^{17}O from $H_2{}^{17}O$ in human erythrocytes. Biochim. Biophys. Acta **385**: 81-87.
32. BRAHM, J. 1982. Diffusional water permeability of human erythrocytes and their ghosts. J. Gen. Physiol. **79**: 791-819.
33. JAY, A. W. L. 1975. Geometry of the human erythrocyte. Biophys. J. **15**: 205-222.
34. CANHAM, P. B. & A. C. BURTON. 1968. Distribution of size and shape in populations of normal human red cells. Circ. Res. **22**: 405-422.
35. STECK, T. L. 1974. The organization of proteins in the human red blood cell membrane. J. Cell Biol. **62**: 1-19.
36. MARCHESI, V. T. 1979. Functional proteins of the human red blood cell membrane. Semin. Hematol. **16**: 3-20.
37. JONES, M. N. & J. K. NICKSON. 1981. Monosaccharide transport proteins of the human erythrocyte membrane. Biochim. Biophys. Acta **650**: 1-20.
38. BALDWIN, S. A., J. M. BALDWIN & G. E. LIENHARD. 1982. Monosaccharide transporter of the human erythrocyte. Characterization of an improved preparation. Biochemistry **21**: 3836-3842.
39. ENGELMAN, D. M. 1969. Surface area per lipid molecule in the intact membrane of the human red cell. Nature **223**: 1279-1280.
40. FETTIPLACE, R. 1978. The influence of the lipid on the water permeability of artificial membranes. Biochim. Biophys. Acta **513**: 1-10.
41. FINKELSTEIN, A. 1976. Water and nonelectrolyte permeability of lipid bilayer membranes. J. Gen. Physiol. **68**: 127-135.
42. DIX, J. A. & A. K. SOLOMON. 1981. Permeability of red cell lipids to water. Biophys. J. **33**: 46a.
43. LONGUET-HIGGINS, H. C. & G. AUSTIN. 1966. The kinetics of osmotic transport through pores of molecular dimensions. Biophys. J. **6**: 217-224.
44. CHIEN, D. Y. & R. I. MACEY. 1977. Diffusional water permeability of red cells. Biochim. Biophys. Acta **464**: 45-52.
45. HARAN, N. & M. SHPORER. 1976. Study of water permeability through phospholipid vesicle membranes by ^{17}O NMR. Biochim. Biophys. Acta **426**: 638-646.
46. GARY-BOBO, C. M. & A. K. SOLOMON. 1971. Effect of geometrical and chemical constraints on water flux across artificial membranes. J. Gen. Physiol. **57**: 610-622.
47. CHASAN, B. & A. K. SOLOMON. 1979. The reflection coefficient of urea in the human red cell: Effect of pCMBS. The Physiologist **22**: 19.
48. POZNANSKY, M., S. TONG, P. C. WHITE, J. M. MILGRAM & A. K. SOLOMON. 1976. Nonelectrolyte diffusion across lipid bilayer systems. J. Gen. Physiol. **67**: 45-66.
49. GALLUCCI, E., S. MICELLI & C. LIPPE. 1975. Effect of cholesterol on the nonelectrolyte permeability of planar lecithin membranes. Nature **255**: 722-723.
50. RAO, A. & R. A. F. REITHMEIER. 1979. Reactive sulfhydryl groups of the band 3 polypeptide from human erythrocyte membranes. Location in the primary structure. J. Biol. Chem. **254**: 6144-6150.
51. KNAUF, P. A. & A. ROTHSTEIN. 1971. Chemical modification of membranes. II. Permeation paths for sulfhydryl agents. J. Gen. Physiol. **58**: 211-223.
52. RAO, A. 1978. The reactive sulfhydryl groups of the band 3 polypeptide of the human erythrocyte membrane. Ph.D. Thesis. Harvard University. Cambridge, Mass.
53. STECK, T. L., J. J. KOZIARZ, M. K. SINGH, G. REDDY & H. KOHLER. 1978. Preparation and analysis of seven major, topographically defined fragments of band 3, the pre-

dominant transmembrane polypeptide of human erythrocyte membranes. Biochemistry **17**: 1216–1222.

54. Ramjeesingh, M., A. Gaarn & A. Rothstein. 1980. The location of a disulfonic stilbene binding site in band 3, the anion transport protein of the red blood cell membrane. Biochim. Biophys. Acta **599**: 127–139.
55. Jennings, M. L. & H. Passow. 1979. Anion transport across the erythrocyte membrane, in situ proteolysis of band 3 protein, and cross-linking of proteolytic fragments by H_2DIDS. Biochim. Biophys. Acta **554**: 498–519.
56. Jennings, M. L. & M. F. Adams. 1981. Modification by papain of the structure and function of band 3, the erythrocyte anion transport protein. Biochemistry **20**: 7118–7123.
57. Rothstein, A. & M. Ramjeesingh. 1982. The red cell band 3 protein: its role in anion transport. Proc. R. Soc. London Ser. B **299**: 497–507.
58. DuPre, A. M. & A. Rothstein. 1981. Inhibition of anion transport associated with chymotryptic cleavages of red cell band 3 protein. Biochim. Biophys. Acta **646**: 471–478.
59. Ramjeesingh, M., A. Gaarn & A. Rothstein. 1981. The sulfhydryl groups of the 35,000-dalton C-terminal segment of band 3 are located in a 9000-dalton fragment produced by chymotrypsin treatment of red cell ghosts. J. Bioenerg. Biomemb. **13**: 411–423.
60. Vansteveninck, J., R. I. Weed & A. Rothstein. 1965. Localization of erythrocyte membrane sulfhydryl groups essential for glucose transport. J. Gen. Physiol. **48**: 617–632.
61. Batt, E. R., R. E. Abbott & D. Schachter. 1976. Impermeant maleimides. Identification of an exofacial component of the human erythrocyte hexose transport mechanism. J. Biol. Chem. **251**: 7184–7190.
62. Kleinfeld, A. M., M. F. Lukacovic, E. D. Matayoshi & P. Holloway. 1982. Conformation of membrane proteins determined from the spatial distribution of tryptophan. Biophys. J. **37**: 146a.
63. Lukacovic, M. F., A. S. Verkman, J. A. Dix, M. Tinklepaugh & A. K. Solomon. 1982. p-Chloromercuribenzene sulfonate (pCMBS) interaction with band 3 in red cell membranes. Biophys. J. **37**: 215a.
64. Lukacovic, M. F., M. B. Feinstein, R. I. Sha'afi & S. Perrie. 1981. Purification of stabilized band 3 protein of the human erythrocyte membrane and its reconstitution into liposomes. Biochemistry **20**: 3145–3151.
65. Rao, A., P. Martin, R. A. F. Reithmeier & L. C. Cantley. 1979. Location of the stilbenedisulfonate binding site of the human erythrocyte anion-exchange system by resonance energy transfer. Biochemistry **18**: 4505–4516.
66. Dix, J. A., A. S. Verkman, A. K. Solomon & L. C. Cantley. 1979. Human erythrocyte anion exchange site characterized using a fluorescent probe. Nature **282**: 520–522.
67. Verkman, A. S., J. A. Dix & A. K. Solomon. 1983. Anion transport inhibitor binding to band 3 in red blood cell membranes. J. Gen. Physiol. **81**: 421–449.
68. Kleinfeld, A. M., E. D. Matayoshi & A. K. Solomon. 1980. Energy transfer determination of the tryptophan distribution of band 3: a new approach to the study of membrane protein function. J. Supramol. Struct. Suppl. **4**: 68.
69. Snow, J. W., J. F. Brandts & P. S. Low. 1978. The effects of anion transport inhibitors on structural transitions in erythrocyte membranes. Biochim. Biophys. Acta **512**: 579–591.
70. Deuticke, B. 1977. Properties and structural basis of simple diffusion pathways in the erythrocyte membrane. Rev. Physiol. Biochem. Pharmacol. **78**: 1–97.
71. Passow, H., L. Kampmann, H. Fasold, M. Jennings & S. Lepke. 1980. Mediation of anion transport across the red blood cell membrane by means of conformational changes of the band 3 protein. *In* Membrane Transport in Erythrocytes, A. Benzon Symposium 14. U. V. Lassen, H. H. Ussing & J. O. Wieth, Eds. : 345–372. Munksgaard, Copenhagen.
72. Passow, H., H. Fasold, E. M. Gartner, B. Legrum, W. Ruffing & L. Zaki. 1980. Anion transport across the red blood cell membrane and the conformation of the protein in band 3. Ann. N.Y. Acad. Sci. **341**: 361–383.
73. Macara, I. G. & L. C. Cantley. 1981. Mechanism of anion exchange across the red cell

membrane by band 3: interactions between stilbene disulfonate and NAP-taurine binding sites. Biochemistry **20**: 5695–5701.

74. Gunn, R. B. & O. Fröhlich. 1979. Asymmetry in the mechanism for anion exchange in human red blood cell membranes. J. Gen. Physiol. **74**: 351–374.
75. Fröhlich, O. 1982. The external anion binding site of the human erythrocyte anion transporter: DNDS binding and competition with chloride. J. Membr. Biol. **65**: 111–123.
76. Jennings, M. L. 1982. Stoichiometry of a half-turnover of band 3, the chloride transport protein of human erythrocytes. J. Gen. Physiol. **79**: 169–185.
77. Clementi, E., R. Barsotti, J. Fromm & R. O. Watts. 1976. Study of the structure of molecular complexes. XIV. Coordination numbers for selected ion pairs in water. Theoret. Chim. Acta (Berl.) **43**: 101–120.
78. Brahm, J. 1977. Temperature-dependent changes of chloride transport kinetics in human red cells. J. Gen. Physiol. **70**: 283–306.
79. Robinson, R. A. & R. H. Stokes. 1955. *In* Electrolyte Solutions. p. 494. Academic Press. New York.
80. Lepke, S., H. Fasold, M. Pring & H. Passow. 1976. A study of the relationship between inhibition of anion exchange and binding to the red blood cell membrane of DIDS and its dihydro derivative H_2DIDS. J. Membr. Biol. **29**: 147–177.
81. Dalmark, M. & J. O. Wieth. 1972. Temperature dependence of chloride, bromide, iodide, thiocyanate and salicylate transport in human red cells. J. Physiol. **224**: 583–610.
82. Aubert, L. & R. Motais. 1975. Molecular features of organic anion permeability in ox red blood cell. J. Physiol. **246**: 159–179.
83. Giebel, O. & H. Passow. 1960. Die Permeabilität der Erythrocytenmembran für organische Anionen. Pflügers Archiv **271**: 378–388.
84. Sutherland, R. M., A. Rothstein & R. I. Weed. 1967. Erythrocyte membrane sulfhydryl groups and cation permeability. J. Cell. Physiol. **69**: 185–198.
85. Rega, A. F., A. Rothstein & R. I. Weed. 1967. Erythrocyte membrane sulfhydryl groups and the active transport of cations. J. Cell. Physiol. **70**: 45–52.
86. Grinstein, S. & A. Rothstein. 1978. Chemically-induced cation permeability in red cell membrane vesicles. Biochim. Biophys. Acta **508**: 236–245.
87. Solomon, A. K., B. Chasan, J. A. Dix, M. F. Lukacovic, M. R. Toon & A. S. Verkman. 1982. The aqueous pore in the red cell membrane: band 3 as a channel for anions, cations, nonelectrolytes and water. Biophys. J. **37**: 215a.
88. Hille, B. 1975. Ionic selectivity of Na and K channels of nerve membranes. *In* Membranes: A Series of Advances. G. Eisenman, Ed. **3**: 255–323. Marcel Dekker. New York.
89. Läuger, P. & B. Neumcke. 1973. Theoretical analysis of ion conductance in lipid bilayer membranes. *In* Membranes. G. Eisenman, Ed. **2**: 1–59. Marcel Dekker. New York.
90. Mayrand, R. & D. G. Levitt. 1980. Facilitated transport of urea in red cells: saturation kinetics, competitive inhibition by analogs, and asymmetry. Fed. Proc. **39**: 957.
91. Wieth, J. O., J. Funder, R. B. Gunn & J. Brahm. 1974. Passive transport pathways for chloride and urea through the red cell membrane. *In* Comparative Biochemistry and Physiology of Transport. L. Bolis, K. Bloch, S. E. Luria & F. Lynen, Eds. : 317–337. North-Holland Publishing. Amsterdam.
92. Solomon, A. K. & B. Chasan. 1980. Thiourea inhibition of urea permeation into human red cells. Fed. Proc. **39**: 957.
93. Brahm, J. & J. O. Wieth. 1977. Separate pathways for urea and water, and for chloride in chicken erythrocytes. J. Phyiol. **266**: 727–749.
94. Collander, R. 1949. Die Verteilung organischer Verbindungen zwischen Äther und Wasser. Acta Chem. Scandinavica **3**: 717–747.
95. Owen, J. D. & A. K. Solomon. 1972. Control of nonelectrolyte permeability in red cells. Biochim. Biophys. Acta **290**: 414–418.
96. Owen, J. D., M. Steggall & E. M. Eyring. 1974. The effect of phloretin on red cell nonelectrolyte permeability. J. Membr. Biol. **19**: 79–92.

97. JENNINGS, M. L. & A. K. SOLOMON. 1976. Interaction between phloretin and the red blood cell membrane. J. Gen. Physiol. **67**: 381–397.
98. FORMAN, S. A., A. S. VERKMAN, J. A. DIX & A. K. SOLOMON. 1982. Effect of lipid perturbants on red cell band 3 conformational states. Biophys. J. **37**: 216a.
99. SEEMAN, P. 1972. The membrane actions of anesthetics and tranquilizers. Pharmacol. Rev. **24**: 583–655.
100. GUIDOTTI, G. 1977. The structure of intrinsic membrane proteins. J. Supramol. Struct. **7**: 489–497.
101. ENGELMAN, D. M. & T. A. STEITZ. 1981. The spontaneous insertion of proteins into and across membranes: The helical hairpin hypothesis. Cell **23**: 411–422.
102. ENGELMAN, D. M., R. HENDERSON, A. D. MCLACHLAN & B. A. WALLACE. 1980. Path of the polypeptide in bacteriorhodopsin. Proc. Natl. Acad. Sci. USA **77**: 2023–2027.
103. RAMJEESINGH, M., S. GRINSTEIN & A. ROTHSTEIN. 1980. Intrinsic segments of band 3 that are associated with anion transport across red blood cell membranes. J. Membr. Biol. **57**: 95–102.
104. SHA'AFI, R. I., C. M. GARY-BOBO & A. K. SOLOMON. 1971. Permeability of red cell membranes to small hydrophilic and lipophilic solutes. J. Gen. Physiol. **58**: 238–258.
105. SAVITZ, D., V. W. SIDEL & A. K. SOLOMON. 1964. Osmotic properties of human red cells. J. Gen. Physiol. **48**: 79–94.
106. MORGAN, J. & B. E. WARREN. 1938. X-ray analysis of the structure of water. J. Chem. Phys. **6**: 666–673.
107. DODGE, J. T., C. MITCHELL & D. J. HANAHAN. 1963. The preparation and chemical characteristics of hemoglobin-free ghosts of human erythrocytes. Arch. Biochem. Biophys. **100**: 119–130.
108. POZNANSKY, M. & A. K. SOLOMON. 1972. Regulation of human red cell volume by linked cation fluxes. J. Membr. Biol. **10**: 259–266.
109. VERKMAN, A. S. & J. A. DIX. 1981. Theoretical interpretation of the kinetics of complex biological systems: Some useful numerical approximations. Biophys. J. **33**: 188a.
110. SHA'AFI, R. I., G. T. RICH., D. C. MIKULECKY & A. K. SOLOMON. 1970. Determination of urea permeability in red cells by minimum method. J. Gen. Physiol. **55**: 427–450.

ARACHIDONIC ACID, LEUKOTRIENE B_4, AND NEUTROPHIL ACTIVATION*

P. H. Naccache and R. I. Sha'afi

Departments of Pathology and Physiology
University of Connecticut Health Center
Farmington, Connecticut 06032

Introduction

Plasma membranes play an essential role in the establishment and the maintenance of the transmembrane conditions necessary for the transmission of biological signals. Originally much of the work directed towards the elucidation of the mechanism of signal transduction focused on the protein components of membranes, since it was thought that in these lie the molecular basis for hormone recognition, transmembrane signaling, and gated or controlled ionic channels.[1-4] It has become increasingly clear, however, that the lipid moieties of the plasma membrane play, in concert with or independent of the protein constituents, an essential role in many of these biological functions.[5] The modulation of the reactivity of several membrane enzymes, such as various ATPases, and of the transport activity of ion carriers by various phospholipids has been well documented.

The hormone-like actions of arachidonic acid metabolites have been recognized somewhat more recently. Prostaglandins represent probably the best known such family of components. Their effects, often mediated by the alteration of the levels of cyclic nucleotides, resemble significantly more those of classical hormones than does the modulation of enzymatic activities by phospholipids.

A new metabolic pathway for arachidonic acid has recently been discovered and characterized.[6,7] This pathway is initiated by one or more of several lipoxygenases differing in positional specificity. A whole family of compounds with potent biological activities are thus generated, the best known being the various leukotrienes. From the limited information gathered in the few years since their discovery, leukotrienes have been shown to be of critical importance in various immediate hypersensitivity reactions and to activate mast cells, vascular and pulmonary smooth muscle cells, and several types of leukocytes.

The rest of the presentation will be devoted to the description of a cellular model that has been used to study the mechanism of action of arachidonic acid.

Cellular Model for the Study of the Mode of Action of Arachidonic Acid

Polymorphonuclear leukocytes (or neutrophils) present an excellent opportunity for such a study. Several neutrophil stimuli of probable physiological and pathological significance are known and available. The functions elicited are well defined and easy to assay. Finally, neutrophils are known to respond to the addition of exogenous arachidonic acid and to metabolize it through one or more metabolic pathways depending on the species (see Klebanoff and Clark[8] for an exhaustive survey of the neutrophils).

* Supported in part by National Institutes of Health grants AI-13734, AI-09648, and AM-31000.

0077-8923/83/0414-0125$01.75/0 © 1983, NYAS

Neutrophils present the first line of defense against foreign and pathogenic elements. As such they have to perform several functions. Neutrophils detect, move towards, and accumulate at sites of injury or infection. This phenomenon, elicited by the detection of soluble factors, is termed directed locomotion or chemotaxis. Most neutrophil stimuli are chemotactic factors, i.e. they cause neutrophils to move up a concentration gradient of the stimulus.

Once at the site of infection, neutrophils ingest or phagocytize the desired particles. This process is followed by the discharge of the neutrophils' various lysosomes and granules into the phagocytic vacuole thereby promoting the degradation and neutralization of the engulfed particle or organism.

Chemotaxis, phagocytosis, degranulation, and aggregation with the attendant changes in cell shape,[9] all depend on the mechanical displacement of part of or the whole cell. Mechano-chemical coupling is thus an essential component of each of these functions.

It is partly on the basis of the analogy to excitation-contraction coupling in skeletal muscle, implicit in the previous statement, that efforts have been made over the past few years to define the sequence of events that leads from the recognition of chemotactic factors by neutrophils to the expression of the desired function (FIGURE 1).

The description by Schiffman *et al.*[10] of the chemokinetic and chemotactic activity of synthetic formyl-oligopeptides has led to the identification of plasma membrane–located specific binding sites on neutrophils and monocytes.[11–13] These receptors behave much the same as classical peptide hormone receptors as far as their physical characteristics thus allowing the definition of one or more[14,15] sets of binding sites with distinct affinities. In addition, these binding sites, and the functions elicited upon their occupation, display the characteristic down-regulation or desensitization behavior upon a prior exposure to the same stimuli[16–19] (see also Becker *et al.*[20] for a recent summary of the data on down-regulation in neutrophils). Work is currently in progress in several laboratories on the understanding of the physiologi-

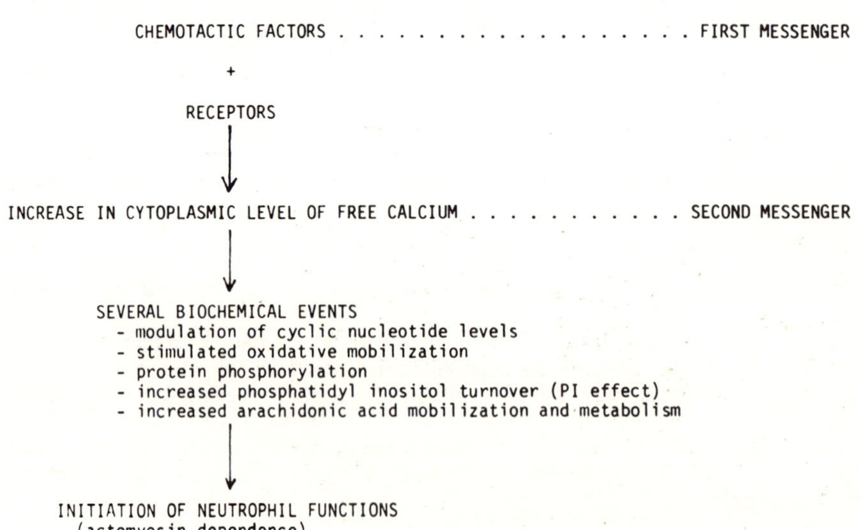

FIGURE 1. Schematic diagram of neutrophil activation.

cal regulation of the binding affinities and on the molecular identification of the membrane components there involved.[21,22]

It is generally agreed that the neutrophil's second messenger function is fulfilled by calcium.[23] The evidence concerning the modulation of the intracellular levels of calcium will be summarized later. The postulated central role of calcium in neutrophil activation is in keeping with these cells' requirement for mechano-chemical transduction for the performance of their functions. The identification of actin, myosin, and several regulatory proteins in most non-muscle cells,[24,25] including the neutrophils,[26,27] reinforces this concept.

Receptor occupation by chemotactic factors activates several biochemical events. These include rapid changes in cation (Na^+ and Ca^+) movements,[28-31] rapid changes in the plasma membrane potential,[32-36] transient rises in cyclic AMP[37-40] and possibly cyclic GMP,[41] the stimulation of the oxidative metabolism of the neutrophils,[42-44] the phosphorylation of several proteins,[45] and finally lipid changes, such as increased phosphatidylinositol turnover (the PI effect),[46,47] phospholipid methylation,[47,48] and increased arachidonic acid mobilization and metabolism.[48,50-54] It is upon the latter two events, and their possible relationship to the modulation of the intracellular levels of calcium (the second messenger), that we would like to focus now.

Arachidonic Acid Metabolism

Two critical features of the handling of arachidonic acid by mammalian cells underlie its modulatory role of various cellular functions (Figure 2). Under resting conditions most animal cells contain little if any free arachidonic acid. Relatively large changes in the concentration of free arachidonic acid can thus be accomplished

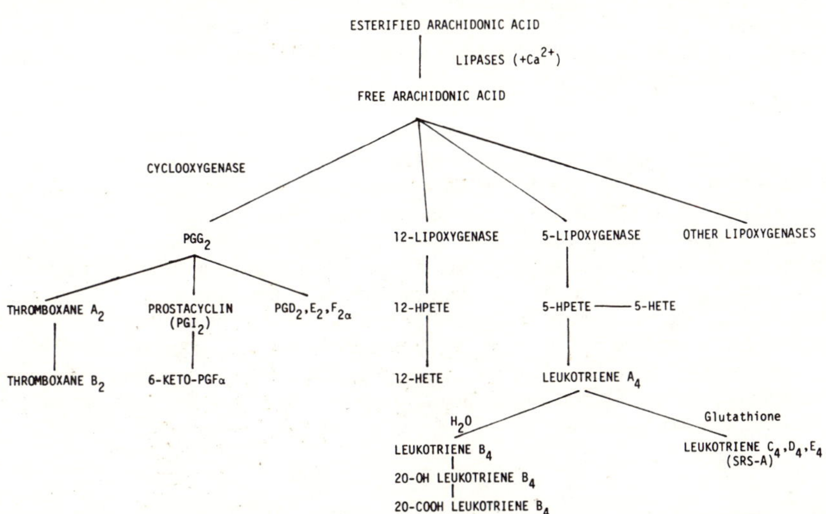

Figure 2. Diagrammatic representation of the major metabolic pathways of arachidonic acid. Emphasis was placed on those pathways leading to the generation of compounds with known biological activities. The interactions between lipoxygenases and cyclooxygenase and their products have been omitted for simplicity.

upon the liberation of small amounts of previously esterified fatty acid. The biochemical mechanisms related to the stimuli-induced liberation of arachidonic acid are still somewhat controversial although the involvement of one or more lipases (phospholipase A_2 and C and diglyceride lipase in particular) is generally accepted.[55] Once released, arachidonic acid is rapidly converted to sets of biologically active compounds (prostaglandins, thromboxanes, and leukotrienes) and thus, the liberation of small quantities of arachidonic acid results in the generation of large amounts of biological activity.

Arachidonic acid metabolism generates three families of compounds with biological activities. The synthesis of the prostaglandins and the thromboxanes is initiated by the cyclo-oxygenase and the synthesis of the leukotrienes by the lipoxygenases. The relative importance of each of these pathways depends on the cell source. In some cells, such as the platelets, arachidonic acid is metabolized by the three pathways, while in other cells, such as the neutrophils, the metabolism of arachidonic acid proceeds predominantly through the lipoxygenase pathway.

Several lipoxygenases with differing positional specificities have been described. Here again the exact mix of these various enzymes depends on the cellular source, the platelet exhibiting predominantly the 12-lipoxygenase and the neutrophils, both the 5- and the 15-lipoxygenases. In each case the corresponding hydroperoxy derivative of arachidonic acid is formed. The latter can either be spontaneously hydrolyzed to form the corresponding hydroxy eicosatetraenoic acid (HETE) or, in the case of the 5-lipoxygenase, be enzymatically converted to an unstable epoxide, leukotriene A_4, that reacts rapidly with water to form leukotriene B_4, a dihydroxy derivative of arachidonic acid or, in the presence of cysteine-containing peptides, to leukotriene C_4, D_4, and E_4.[6,7] Leukotriene B_4 has recently been shown to be ω-oxidized to form 20-OH leukotriene B_4, the latter being further metabolized to the dicarboxylic acid 20-COOH-leukotriene B_4.[54]

The interrelation between the various lipoxygenases and their products is currently under active investigation. Double dioxygenation of arachidonic acid through the 5- and 12-lipoxygenase and the 5- and 15-lipoxygenase have thus been demonstrated.[56] In addition, 5-HPETE has been shown to increase the release of arachidonic acid induced by chemotactic factors in the human promyelocytic leukemia cell line HL60.[57]

The significance of the various leukotrienes rests in their profound effects both on the cells of origin and on target cells (FIGURE 3).[6,7,58,61] Leukotrienes C_4, D_4, and E_4 have thus been shown to be the active constituents of slow-reacting substance of anaphylaxis (SRS-A) and as such are potent bronchoconstrictors (leukotrienes are about a 1,000-fold more active than histamine in this respect) and vascular permeability-increasing factors. Leukotriene B_4 is one of the most active neutrophil stimuli, inducing, among others, at nanomolar and subnanomolar concentrations, chemotactic, aggregatory, and degranulation responses from neutrophils of several species. The complexity of the *in vivo* effects of leukotrienes and of their modulation has been highlighted by the description of synergistic effects on plasma exudation between leukotrienes and prostaglandins and by the opposing effects of the latter on bronchial constriction.

MECHANISM OF ACTION OF LEUKOTRIENE B_4

Understanding the basis for the actions of these various lipid stimuli and others, such as the recently characterized platelet-activating factor, has thus become a ques-

LEUKOTRIENE C_4, D_4 — SMOOTH MUSCLE CONTRACTION

CONTRACT
- -GUINEA PIG ILEUM
- -HUMAN AND GUINEA PIG BRONCHI (SMALL AIRWAYS)
- -HUMAN PULMONARY ARTERY AND VEIN

DILATE SMALL VESSELS FROM SKIN (LTD_4)

INCREASE MUCUS GLYCOPROTEIN SYNTHESIS AND DECREASE RATE OF MUCUS CLEARANCE FROM ASTHMATIC PATIENTS

INCREASE PLASMA LEAKAGE (ENHANCED BY PGE_2)

LEUKOTRIENE B_4 — NEUTROPHIL STIMULUS

INDUCES
- -CHEMOTAXIS
- -AGGREGATION
- -DEGRANULATION
- -EXPRESSION OF SURFACE $C3_b$ RECEPTORS

INCREASES PLASMA LEAKAGE SYNERGISTICALLY WITH PGE_2 AND NEUTROPHILS

FIGURE 3. A summary of the major biological activities of the various leukotrienes.

tion central to cell physiology and to the clinical status of immediate hypersensitivity, asthma, and inflammation.

The majority of the neutrophil functions are, as already pointed out, contractile-dependent events. The bronchoconstrictive and contractile responses to SRS-A (and to the leukotrienes) are similarly presumably dependent on calcium-activated events.

In part, on the basis of these considerations, we have recently been detailing the effects of one leukotriene, leukotriene B_4, on calcium mobilization in neutrophils, one of the tissues of origin of leukotriene B_4 and its apparently major target.

These investigations, and the concurrent functional studies (chemotaxis, degranulation, and aggregation) have proceeded along two main lines: (1) the direct measurement of the effects of exogenously added arachidonic acid, the precursor of the leukotrienes, or of individual, purified lipoxygenase metabolites and (2) a pharmacological approach designed to determine the effects of inhibitors of arachidonic acid mobilization and metabolism on stimulated neutrophil functions.

A brief presentation of the mechanism of calcium mobilization in rabbit neutrophils is warranted before proceeding to a description of the effects of arachidonic acid and leukotriene B_4.

Chemotactic factors, such as the synthetic peptide formyl-methionyl-leucyl-phenylalanine (f-Met-Leu-Phe) and the small molecular weight fragment of the fifth component of complement (C5a), have been shown to increase rapidly the neutrophils' level of exchangeable calcium.[28-31,62] This is manifested by an increase in the steady-state level of cell-associated ^{45}Ca in the presence of concentrations of extracellular calcium in excess of 50 μM (FIGURE 4). This increase is dose-dependent and occurs at physiologically relevant concentrations of the chemotactic factors.

The additional calcuim required for the increase in the neutrophils' level of ex-

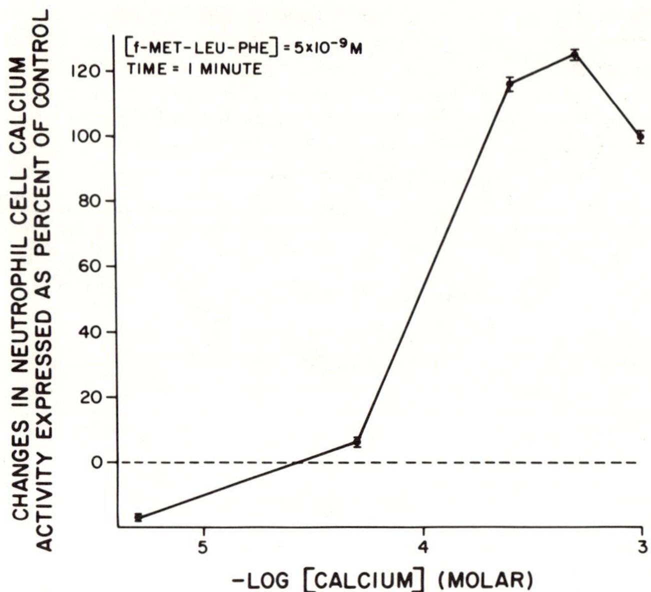

FIGURE 4. The extracellular calcium dependence of the f-Met-Leu-Phe–induced changes in the steady-state levels of the neutrophils' calcium-specific activity. This graph is derived from data presented in Petroski *et al.*[29] in which the experimental protocols are detailed.

changeable calcium is derived from two sources. An intracellular source was experimentally detected by measuring one or more of three parameters: (1) the steady-state level of cell-associated ^{45}Ca in the presence of low (< 50 μM) concentrations of extracellular calcium; i.e., in the absence of an inwardly directed electrochemical gradient for calcium, (2) the rate of ^{45}Ca efflux from preloaded cells, and (3) the fluorescent characteristics of chlortetracycline, a fluorescent chelate probe sensitive to the presence or absence of calcium in its environment.[63–65] An extracellular source manifested indirectly by increases in the initial rate of ^{45}Ca influx, the extracellular calcium dependence of the changes in the steady-state levels of exchangeable calcium, and, finally, by the enhancement of the functional responsiveness of neutrophils that is observed in the presence of calcium as compared to that seen in its absence.

The present working model of chemotactic factor–induced calcium mobilization in neutrophils thus possesses the following characteristics: the binding of the chemotactic factors to their plasma membrane receptors causes the graded release of a previously unexchangeable pool of membrane calcium; subsequent to the release of calcium, and possibly as a consequence of it, the inward plasma membrane permeability to calcium is increased; as a result of the above events, the intracellular concentration of exchangeable, and by extension, free calcium is increased thereby initiating neutrophil responsiveness; and activating the membrane-located, ATPase-driven calcium pump[45,97] responsible for the restoration of the cytoplasmic level of calcium.

It is thus believed that the intracellular redistribution of calcium is one of the earliest events following receptor occupation and that it plays a central and essential role in initiating the various neutrophils functions previously mentioned. In this view, the role of extracellular calcium, and of the chemotactic factor–induced net calcium uptake, is a modulatory or amplifying one. In the presence of extracellular calcium, significantly lower concentrations (1/10 to 1/100) of chemotactic factors are required to achieve the same level of activation than in its absence (see Sha'afi and Naccache[23] for more details about calcium homeostasis in neutrophils).

At submicromolar concentrations exogenous arachidonic acid induces chemotactic, secretory, and aggregatory responses from neutrophils from some species, such as rabbit, but essentially none from others, such as human cells. These species differences are thought to be related to the endogenous metabolic activity of the various cell populations—rabbit peritoneal neutrophils metabolize exogenous arachidonic acid to a significantly greater extent than human peripheral neutrophils, for example.

The characteristics of the rabbit neutrophil responses to arachidonic acid in many ways resemble those responses to the chemotactic factors. Many of the same functions are elicited, e.g., chemotaxis, aggregation, and degranulation. The similarities between the degranulation of rabbit neutrophils induced by f-Met-Leu-Phe and arachidonic acid are particularly striking. Both stimuli require the presence of the fungal metabolite cytochalasin B for the expression of their secretory activity. Secretion proceeds rapidly, being essentially complete by 2–3 minutes, and is significantly enhanced in the presence of extracellular calcium.[66] In addition, the responses to these two stimuli are essentially similarly sensitive to a variety of inhibitors.[69] A similar parallelism between f-Met-Leu-Phe and arachidonic acid–induced neutrophil aggregation has also been noted.[67]

In view of the metabolic pathways available to arachidonic acid and of the species' specificity of its biological effects, the question immediately arose as to whether its neutrophil-directed biological activities were caused by itself or by one or more of its metabolites. The answer to this question was biased towards the latter possibility when it was found that several compounds with antagonist properties towards arachidonic acid metabolism (such as the lipoxygenase inhibitors nordihydroguaiaretic acid, 5,8,11,14-eicosatetraynoic acid, quercetin, and BW 755C) all significantly depressed the responses of the neutrophil to chemotactic factors and to arachidonic acid.[68–71] In addition, a direct correlation has been demonstrated between the inhibition of the arachidonic acid–induced neutrophil degranulation by nordihydroguaiaretic acid and the inhibition of the lipoxygenase activity of these cells.[72]

More direct confirmation of a role for arachidonic acid metabolites in neutrophil responsiveness came from studies with purified lipoxygenase products. The results of these studies demonstrated that several products of the lipoxygenase pathway possessed chemotactic activity towards neutrophils. The activity of the monoHETE as chemoattractants is expressed at concentrations in the range of 0.1–10 micromolar,[73,74] essentially no increase in activity over that of arachidonic acid. Leukotriene B_4, on the other hand, has recently been found to be a potent neutrophil stimulus at nanomolar concentrations.[75–80] Leukotriene B_4 is thus, on a molar basis, at least 100 to 1,000 times more active than arachidonic acid or the monoHETEs. Recent studies with the natural metabolites of leukotriene B_4 have in addition demonstrated that the latter are significantly less active than leukotriene B_4 as neutrophil and smooth muscle agonists.[81,82]

It is thus reasonably clear that most of the agonist properties of arachidonic acid towards neutrophils are attributable to the generation of leukotriene B_4. The formation from endogenous arachidonic acid of 20-OH leukotriene B_4, and by implication, of leukotriene B_4 in human neutrophils,[54] and that of leukotriene B_4 itself from the human promyelocytic leukemia cell line HL60[83] upon stimulation by chemotactic factors has also been recently reported.

Little is known about the mechanism of action of leukotriene B_4. Its activity has

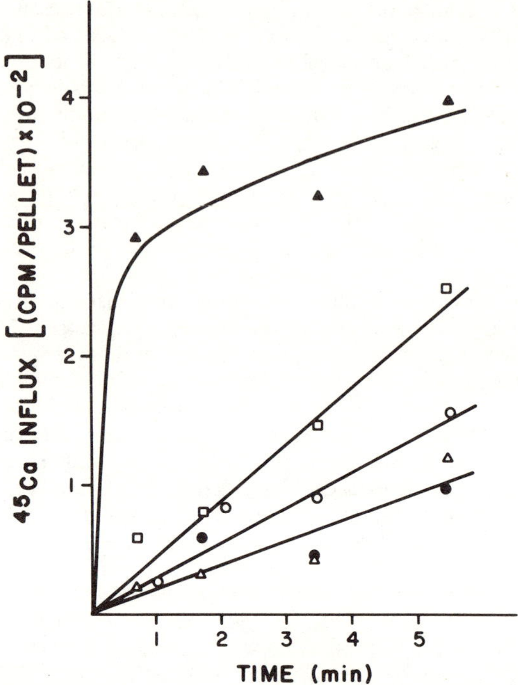

FIGURE 5. Effect of leukotriene B_4 and of some of its stereo isomers on the initial rate of ^{45}Ca influx in rabbit neutrophils. This graph summarizes data previously published in Naccache *et al.*[90,91] Symbols are as follows: filled circles: no additions; filled triangles: leukotriene B_4; open squares: Δ^6-*trans*-12 epi-leukotriene B_4; open triangles: Δ^6-*trans*-leukotriene B_4; open circles: 5(S),12(S)diHETE.

been found to be extremely sensitive to the stereogeometry of the double bonds in the triene structure and of the hydroxyl groups, modifications of the natural configuration resulting in shifts in activity of 50- to 100-fold.[82,85–87] The sensitivity of the biological activity of leukotriene B_4 to even subtle changes in its configuration coupled with competition experiments with analogues and derivatives of leukotriene B_4,[88] have led to the suggestion that the effects of leukotriene B_4 were receptor mediated. The leukotriene B_4 receptors appear to be distinct from those of the formylmethionyl peptides or of C5a.[88] The location and characterization of these putative receptors has just started[89] and remains to be carried out in details.

Leukotriene B_4, at concentrations similar to those required for the expression of its biological activities, rapidly increases the rate of uptake of ^{45}Ca in rabbit neutrophils.[82,90,91] Leukotriene B_4 is at least 100 times more active than any of the mono-HETEs tested. The efficacy of leukotriene B_4 in the calcium uptake assay thus closely parallels its chemotactic, aggregatory, and secretory activities.[82] The effect of leukotriene B_4 on calcium uptake displayed the same stereospecificity as that of its biological effects, i.e. only natural leukotriene B_4 is active (FIGURE 5). In addition, the activity of the leukotriene B_4 decreases as it is further metabolized by the cells (FIGURE 6). The ω-hydroxylation of leukotriene B_4 results in a compound nearly as active as leukotriene B_4 itself as far as its ability to increase ^{45}Ca uptake and to aggre-

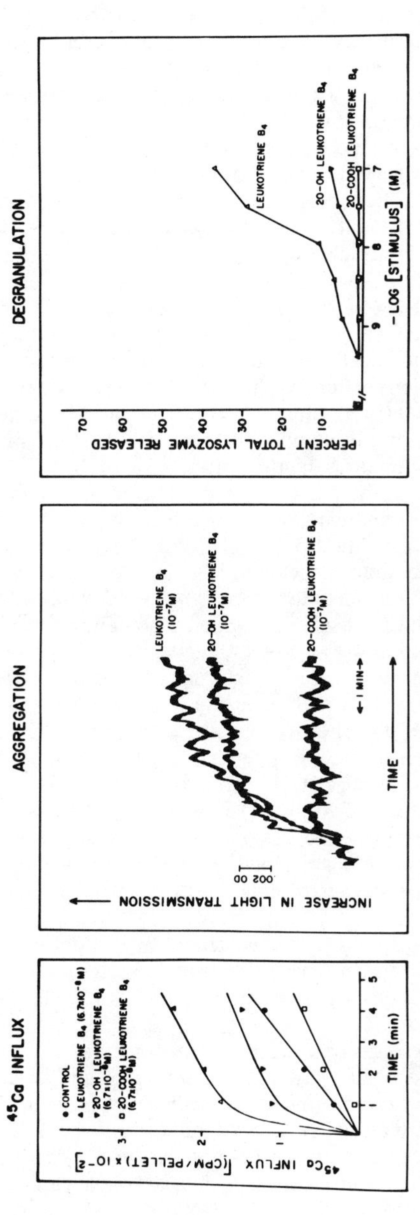

FIGURE 6. Effects of leukotriene B_4, 20-OH-leukotriene B_4, and 20-COOH-leukotriene B_4 on the initial rate of ^{45}Ca influx, and on the aggregation and degranulation of rabbit neutrophils. The experimental protocols are detailed in Naccache *et al.*[82]

gate the cells. 20-OH leukotriene B_4 is however significantly less active than leukotriene B_4 as a neutrophil secretagogue. The dicarboxylic acid 20-COOH-leukotriene B_4 is essentially inactive, up to the concentrations tested, in the calcium uptake and the functional assays.[82] The basis for the differentiation by 20-OH leukotriene B_4 between aggregation and degranulation is not known at present although it might be related to the existence of different thresholds of cytoplasmic calcium for the initiation of different functions—degranulation requiring, according to this hypothesis, higher levels of calcium than aggregation.

The finding that leukotriene B_4 increased the rate of ^{45}Ca uptake and the apparent close parallelism between the ability of the various metabolites to increase the rate of ^{45}Ca uptake and induce several neutrophil functions prompted a detailed examination of the effects of leukotriene B_4 on calcium homeostasis in rabbit neutrophils.[90] These studies were modeled upon the scheme of chemotactic factor–induced calcium mobilization previously outlined.

Leukotriene B_4, as f-Met-Leu-Phe, rapidly increases the neutrophils' levels of exchangeable calcium. This additional calcium appears to be derived from the extracellular medium and an intracellular source.

The evidence for a leukotriene B_4–induced increase in permeability to calcium is, as was the case for f-Met-Leu-Phe, indirect, as it cannot be demonstrated experimentally in the absence of the concurrent redistribution of intracellular calcium. Nevertheless, the rapid increases in the initial rate of uptake of ^{45}Ca and in the steady-state level of ^{45}Ca in the presence of close to physiological concentrations of calcium, together with the enhancing effect of extracellular calcium on the leukotriene B_4–induced neutrophil degranulation,[80] all argue for the existence of a stimulus-induced increase in permeability to calcium.

The redistribution of intracellular calcium is manifested under two sets of experimental conditions. (1) In the absence of added extracellular calcium, leukotriene B_4 causes a transient decrease in the steady-state level of ^{45}Ca. The increase in cytoplasmic concentration of free calcium resulting from the sudden release of a pool of previously unexchangeable calcium, activates the outward-directed calcium pump and produces the observed drop in the steady-state level of ^{45}Ca. The subsequent return to the prestimulation level of ^{45}Ca is a reflection of the reequilibration of the cells' specific activity of ^{45}Ca temporarily lowered by the addition of previously unexchangeable calcium and thus of low specific activity. (2) Leukotriene B_4 rapidly increases the rate of ^{45}Ca efflux from preloaded cells. This parameter, particularly in conjunction with the previously described changes in the steady-state level of ^{45}Ca, has been interpreted in several systems as a reflection of an intracellular redistribution of calcium.

It is thus clearly established that leukotriene B_4 is a potent and multifunctional neutrophil stimulus and it appears probable that its activities are related to its ability to mobilize calcium by essentially the same mechanisms as chemotactic factors such as f-Met-Leu-Phe.

Two interrelated questions arise at this point (FIGURE 7). First, given that the activation of the lipoxygenase pathway is sufficient to initiate several neutrophil functions, is it also a necessary condition for the cell's responsiveness? Secondly, what is the physiological role of leukotriene B_4? Is leukotriene B_4 a "first messenger" and therefore a sufficient, but not necessary condition, designed to amplify the recruitment of neutrophils to the required sites, or is it an intracellular mediator of the effects of chemotactic factors and therefore both a necessary and sufficient condition for neutrophil activation?

Although the evidence available to date does not allow an unambiguous answer

A. INTRACELLULAR MEDIATOR OF THE ACTIONS OF CHEMOTACTIC FACTORS

(A necessary and sufficient condition for neutrophil activation)

B. INTER-CELLULAR ("FIRST MESSENGER") ROLE

(A sufficient but not necessary condition for neutrophil activation)

FIGURE 7. The possible physiological roles of leukotriene B_4 and their causal relationship to neutrophil activation.

to the above two questions, the following lines of evidence argue in favor of an intercellular role for leukotriene B_4.

(1) Leukotriene B_4, when formed upon stimulation by chemotactic factors, is recovered rapidly in the cells' supernatants.[84] An *in vivo* correlate of this is the recovery of leukotriene B_4 from various inflammatory fluids.[92] One might expect a compound produced by cells for an intracellular role not to diffuse so readily to the extracellular fluids.

(2) Exogenous leukotriene B_4 mimics several of the effects of chemotactic factors and acts just about as rapidly as the latter while a delay may be expected for an intracellular mediator to diffuse into the cells. It may be relevant to note that leukotriene B_4 does not stimulate the O_2 consumption and the production of potentially toxic oxygen species in neutrophils.[88] This would be a desirable feature of a compound whose prime function is to facilitate or speed up the movement of cells along healthy surfaces and towards sites of infection.

(3) Finally, it has been shown that neutrophils preincubated with chemotactic factors such as f-Met-Leu-Phe and C5a lose their ability to respond to further stimulation by the same factors and by arachidonic acid and leukotriene B_4. On the other hand, preincubation of the neutrophils with the lipid factors only reduces the ability of the neutrophils to respond to arachidonic acid and to leukotriene B_4. The response to f-Met-Leu-Phe is unaffected.[80,93,94] The direct implication of these results is that f-Met-Leu-Phe is capable of activating neutrophils even when the arachidonic acid-leukotriene B_4 pathway is functionally ineffective (i.e. deactivated).

The ambiguities inherent in the interpretation of the results just summarized should however be kept in mind. It should prove difficult to clearly distinguish between the above two hypotheses concerning the exact role of leukotriene B_4 as most of the criteria used to define first and second messengers are at least experimentally, if not conceptually, related.

The mechanism by which leukotriene B_4 mobilizes calcium in rabbit neutrophils differs in several important features from that of a classical calcium ionophore, such as A23187.

First and perhaps most fundamentally, the effects of leukotriene B_4 are cell specific while A23187 will transport calcium across virtually any lipid bilayer, biological and synthetic.

Secondly, the cell specificity of the effects of leukotriene B_4 together with the extreme sensitivity of the various biological activities of leukotriene B_4 to relatively minor stereochemical modifications, argue strongly in favor of a specific binding site or receptor. A23187 on the other hand is thought to diffuse and partition non-specifically into the hydrophobic environments of the membranes it is added to.

Thirdly, leukotriene B_4, in addition to increasing the rate of uptake of calcium, also enhances the initial rate of uptake of ^{22}Na in rabbit neutrophils.[91] The latter, but not the former effect of leukotriene B_4 is sensitive to the K^+-sparing diuretic amiloride.[95] A23187 exhibits significantly greater selectivity than leukotriene B_4 towards calcium. Its effects are unlikely to be inhibited by amiloride.

Fourthly, the deactivation experiments previously described are difficult to reconcile with the concept of an ionophore. Why should the pretreatment of the cells with a peptide acting through its plasma membrane receptors inhibit leukotriene B_4, were it to be an ionophore, from stimulating ^{45}Ca uptake into neutrophils?

Conclusions

The model that emerges from these considerations is therefore one in which neutrophils first recognize one of the "classical" chemotactic factors, such as f-Met-Leu-Phe and C5a. The latter induce, probably in a Ca^{2+}-dependent manner,[96] the generation and release of leukotriene B_4. The fatty acid will then interact specifically with binding sites or receptors on (or in) neighboring cells. Activating essentially the same events as chemotactic factors, leukotriene B_4 then increases calcium mobilization and cellular responsiveness. The essential function of leukotriene B_4 would therefore be to amplify or stabilize the chemotactic gradients.

The studies presented above represent the initial attempt at defining the mode of action of one leukotriene (leukotriene B_4) in one cell (the neutrophils). Significantly more work is required to characterize the latter in terms less tentative than those presently used. In addition, the mechanism of action of the other leukotrienes (leukotriene C_4, D_4, and E_4) remains essentially totally unknown. The possibility of similarities between the mechanism of action of leukotriene B_4 and that of leukotriene C_4, D_4, and E_4 appears to be worthy of exploration.

References

1. Sonenberg, M. & A. S. Schneider. 1977. *In* Receptors and Recognition. P. Cuatrecasas & M. F. Greaves, Eds. 4: 1–72. Chapman and Hall. London.
2. Catt, K. J. & M. L. DuFau. 1977. Ann. Rev. Physiol. **39**: 529–557.
3. Triggle, D. J. 1980. *In* Membrane Structure and Function. E. E. Bitter, Ed. 3: 2–58. John Wiley and Sons. New York.
4. Rodbell, M. 1980. Nature **284**: 17–22.
5. Loh, H. H. & P. Y. Law. 1980. Ann. Rev. Pharmacol. Toxicol. **20**: 201–234.
6. Samuelsson, B., S. Hammarstrom & P. Borgeat. 1979. *In* Advances in Inflammation Research. G. Weissman, B. Samuelsson & R. Paoletti, Eds. 1: 405–412. Raven Press. New York.
7. Samuelsson, B., S. Hammarstrom, R. C. Murphy & P. Borgeat. 1980. Allergy **35**: 375–381.
8. Klebanoff, S. J. & R. A. Clark. 1978. The Neutrophil: Function and Clinical Disorders. North Holland Publishing Company. Amsterdam.
9. Craddock, P. R., D. Hammerschmidt, J. G. White, A. P. Dalmasso & H. S. Jacob. 1977. J. Clin. Invest. **60**: 260–264.
10. Schiffmann, E., B. A. Corcoran & S. M. Wahl. 1975. Proc. Natl. Acad. Sci. USA **72**: 1059–1062.
11. Aswanikumar, S., B. A. Corcoran, E. Schiffmann, A. R. Day, R. J. Freer, H. J. Showell & E. L. Becker. 1977. Biochem. Biophys. Res. Commun. **74**: 810–817.
12. Williams, L. T., R. Snyderman, M. C. Pike & R. J. Lefkowitz. 1977. Proc. Natl. Acad. Sci. USA **74**: 1204–1208.
13. Sha'afi, R. I., K. Williams, M. C. Wacholtz & E. L. Becker. 1978. FEBS Lett. **91**: 305–309.

14. Coo, C., R. T. Lefkowitz & R. Snyderman. 1982. Biochem. Biophys. Res. Commun. **106**: 442–444.
15. Mackin, W. M., C. K. Huang & E. L. Becker. 1982. J. Immunol. (In press.)
16. Ward, P. A. & E. L. Becker. 1968. J. Exp. Med. **127**: 693–709.
17. Sullivan, S. J. & S. H. Zigmond. 1980. J. Cell. Biol. **85**: 703–711.
18. Vitkauskas, G., H. J. Showell & E. L. Becker. 1980. Mol. Immunol. **17**: 171–180.
19. Sha'afi, R. I., T. F. P. Molski, P. Borgeat & P. H. Naccache. 1981. Biochem. Biophys. Res. Commun. **103**: 766–773.
20. Becker, E. L., P. H. Naccache, R. W. Walenga & H. J. Showell. 1981. Lymphokine Rep. **4**: 297–334.
21. Niedel, J. 1981. J. Biol. Chem. **256**: 9295–9299.
22. Goetzl, E. J., D. W. Foster & D. W. Goldman. 1981. Biochemistry **20**: 5717–5722.
23. Sha'afi, R. I. & P. H. Naccache. 1981. Adv. Inflammation Res. **2**: 115–148.
24. Goldman, R. D., A. Milsted, J. A. Schloss, J. Starger & M. J. Yerna. 1979. Annu. Rev. Physiol. **41**: 703–722.
25. Korn, E. D. 1978. Proc. Natl. Acad. Sci. USA **75**: 588–599.
26. Stossel, T. P. & T. D. Pollard. 1973. J. Biol. Chem. **248**: 8288–8294.
27. Tatsumi, N., N. Shibata, Y. Okamura, K. Takeuchi & N. Senda. 1973. Biochim. Biophys. Acta **305**: 433–444.
28. Naccache, P. H., H. J. Showell, E. L. Becker & R. I. Sha'afi. 1977. J. Cell Biol. **73**: 428–444.
29. Petroski, R. J., P. H. Naccache, E. L. Becker & R. I. Sha'afi. 1977. Am. J. Physiol. **237**: C43–C49.
30. Gallin, J. I. & A. S. Rosenthal. 1974. J. Cell Biol. **62**: 594–609.
31. Simchowitz, L. & I. Spilberg. 1979. J. Lab. Clin. Med. **94**: 403–413.
32. Simchowitz, L., I Spilberg & P. DeWeer. 1982. J. Gen. Physiol. **79**: 453–479.
33. Korchak, H. M. & G. Weissmann. 1978. Proc. Natl. Acad Sci. USA **75**:3818–3822.
34. Gallin, E. K. & J. I. Gallin. 1974. J. Cell. Biol. **75**: 277–289.
35. Jones, G. S., K. VanDyke & V. Castranova. 1981. J. Cell. Physiol. **106**: 75–83.
36. DosReis, G. A. & G. M. Oliveira-Gastro. 1977. Biochim. Biophys. Acta **469**: 257–263.
37. Jackowski, S. & R. I. Sha'afi. 1979. Molel. Pharmacol. **16**: 473–481.
38. Simchowitz, L., L. C. Fishbein, I. Spilberg & J. P. Atkinson. 1980. J. Immunol. **124**: 1482–1491.
39. Keller, H. U., G. Gerish & J. H. Wessler. 1979. Cell Biol. Intl. Rep. **3**: 759–765.
40. Smolen, J. E., H. M. Korchak & G. Weissmann. 1980. J. Clin. Invest. **65**: 1077–1085.
41. Hatch, G. E., W. K. Nichols & H. R. Hill. 1977. J. Immunol. **119**: 450–456.
42. Schell-Frederick, E. 1974. FEBS Lett. **48**: 37–40.
43. Romeo, D., G. Zabucchi & F. Rossi. 1973. Nature **243**: 111.
44. Badwey, J. A. & M. L. Karnovsky. 1980. Annu. Rev. Biochem. **49**: 695–726.
45. Schneider, C., C. Mottolo & D. Romeo. 1979. Biochem. J. **182**: 655–660.
46. Bennett, J. P., S. Cockroft & B. D. Gomperts. 1980. Biochim. Biophys. Acta **601**: 584–591.
47. Rubin, R. P., L. E. Sink, M. P. Schrey, A. R. Day, C. S. Liao & R. J. Freer. 1980. Biochem. Biophys. Res. Commun. **90**: 1364–1370.
48. Hirata, F., B. A. Corcoran, K. Venkatasubramanian, E. Schiffman & J. Axelrod. 1979. Proc. Natl. Acad. Sci. USA **76**: 2640–2643.
49. Pike, M. C. & R. Snyderman. 1981. J. Cell Biol. **91**: 221–226.
50. Borgeat, P. & B. Samuelsson. 1976. J. Biol. Chem. **251**: 7816–7820.
51. Borgeat, P. & B. Samuelsson. 1979. Proc. Natl. Acad. Sci. USA **76**: 3213–3217.
52. Borgeat, P. & B. Samuelsson. 1979. J. Biol. Chem. **254**: 2643–2646.
53. Bokoch, G. M. & P. W. Reed. 1980. J. Biol. Chem. **255**: 10223–10226.
54. Jubiz, W., O. Radmark, C. Malmsten, G. Hansson, J. A. Lindgren, J. Palmblad, A-M. Uden & B. Samuelsson. 1982. J. Biol. Chem. **257**: 6106–6110.
55. Irvine, R. F. 1982. Biochem. J. **204**: 3–16.
56. Borgeat, P., S. Picard, P. Vallerand & P. Sirois. 1981. Prostagland. Med. **6**: 557–570.
57. Siegel, M. I., R. T. McConnell, R. W. Bonser & P. Cuatrecasas. 1982. Biochem. Biophys. Res. Commun. **104**: 874–881.
58. Goetzl, E. J. 1981. Med. Clin. North Am. **65**: 809–828.

59. Spannhake, E. W., A. L. Littyman & P. J. Kadowitz. 1981. Prostagland. **22**: 1013-1026.
60. O'Driscoll, B. R. C. & A. B. Kay. 1982. Thorax **37**: 241-245.
61. Smith, M. J. H. 1981. Gen. Pharmacol. **12**: 211-216.
62. Boucek, M. M. & R. Synderman. 1976. Science **193**: 905-907.
63. Caswell, A. H. 1972. J. Membr. Biol. **7**: 345-364.
64. Naccache, P. H., M. Volpi, H. J. Showell, E. L. Becker & R. I. Sha'afi. 1979. Science **203**: 461-463.
65. Naccache, P. H., H. J. Showell, E. L. Becker & R. I. Sha'afi. 1980. J. Cell Biol. **83**: 179-186.
66. Naccache, P. H., H. J. Showell, E. L. Becker & R. I. Sha'afi. 1979. Biochem. Biophys. Res. Commun. **89**: 1224-1230.
67. O'Flaherty, J. T., H. J. Showell, E. L. Becker & P. A. Ward. 1979. Inflammation **3**: 431-436.
68. Showell, H. J., P. H. Naccache, R. I. Sha'afi & E. L. Becker. 1980. Life Sci. **27**: 421-426.
69. Showell, H. J., P. H. Naccache, R. W. Walenga, M. Daleki, M. B. Feinstein, R. I. Sha'afi & E. L. Becker. 1981. J. Reticuloendothel. Soc. **30**: 167-181.
70. Goetzl, E. J. 1980. Immunol. **40**: 709-719.
71. Bokoch, G. M. & P. W. Reed. 1979. Biochem. Biophys. Res. Commun. **90**: 481-487.
72. Walenga, R. W., H. J. Showell, M. B. Feinstein & E. L. Becker. 1980. Life Sci. **27**: 1047-1053.
73. Turner, S. R., J. A. Campbell & W. S. Lynn. 1975. J. Exp. Med. **141**: 1437-1441.
74. Goetzl, E. J. & F. F. Sun. 1979. J. Exp. Med. **150**: 406-411.
75. Ford-Hutchinson, A. W., M. A. Bray, M. V. Doig, M. E. Shipley & M. J. H. Smith. 1980. Nature **286**: 264-265
76. Goetzl, E. J. & W. C. Pickett. 1980. J. Immunol. **125**: 1789-1791.
77. Palmer, R. M. J., R. J. Stepney, G. A. Higgs & K. E. Eakins. 1980. Prostagland. **20**: 411-418.
78. Rae, S. A. & M. J. H. Smith. 1981. J. Pharmacol. **33**: 616-617.
79. Bokoch, G. M. & P. W. Reed. 1981. J. Biol. Chem. **256**: 5317-5320.
80. Showell, H. J., P. H. Naccache, P. Borgeat, S. Picard, P. Vallerand, E. L. Becker & R. I. Sha'afi. 1982. J. Immunol. **128**: 811-816.
81. Feinmark, S. J., J. A. Lindgren, H. E. Claesson, C. Malmsten & B. Samuelsson. 1982. FEBS Lett. **136**: 141-144.
82. Naccache, P. H., T. F. P. Molski, E. L. Becker, P. Borgeat, S. Picard, P. Vallerand & R. I. Sha'afi. 1982. J. Biol. Chem. **257**: 8608-8611.
83. Bonser, R. W., M. I. Siegel, R. T. McConnell & P. Cuatrecasas. 1981. Biochem. Biophys. Res. Commun. **98**: 614-620.
84. Bonser, R. W., M. I. Siegel, S. M. Chung, R. T. McConnell & P. Cuatrecasas. 1981. Biochemistry **20**: 5297-5301.
85. Lewis, R. A., E. J. Goetzl, J. M. Drazen, N. A. Sofer, K. Frank-Austen & E. J. Corey. 1981. J. Exp. Med. **154**: 1243-1248.
86. Ford-Hutchinson, A. W., M. A. Bray, F. M. Cunningham, E. M. Davidson & M. J. H. Smith. 1981. Prostagland. **121**: 143-152.
87. Goetzl, E. J. & W. C. Pickett. 1981. J. Exp. Med. **153**: 482-487.
88. Goetzl, E. J., D. W. Goldman, P. H. Naccache, R. I. Sha'afi & W. C. Pickett. 1982. *In* Advances in Prostaglandins, Thromboxane and Leukotriene Research. B. Samuelsson & R. Paoetti, Eds. **9**: 273-283. Raven Press. New York.
89. Kreisle, R. A. & C. W. Parker. 1982. Fed. Proc. **41**: 373.
90. Naccache, P. H., P. Bogeat, E. J. Goetzl & R. I. Sha'afi. 1981. J. Clin. Invest. **64**: 1584-1587.
91. Sha'afi, R. I., P. H. Naccache, T. F. P. Molski, P. Borgeat & E. J. Goetzl. 1981. J. Cell. Physiol. **108**: 401-408.
92. Klickstein, L. B., C. Shapleigh & E. J. Goetzl. 1980. J. Clin. Invest. **66**: 1166-1170.
93. Sha'afi, R. I., T. F. P. Molski, P. Borgeat & P. H. Naccache. 1981. Biochem. Biophys. Res. Commun. **103**: 766-773.

94. O'Flaherty, J. T., R. J. Wykle, C. E. McCall, T. B. Shewmake, C. J. Lees & M. Thomas. 1981. Biochem. Biophys. Res. Commun. **101**: 1290–1296.
95. Sha'afi, R. I., T. F. P. Molski & P. H. Naccache. 1981. Biochem. Biophys. Res. Commun. **90**: 1271–1276.
96. Jakschik, B. A., F. F. Sun, L. Lee-Hank & M. M. Steinhoff. 1980. Biochem. Biophys. Res. Commun. **95**: 103–110.
97. Jackowski, S., K. Petro & R. I. Sha'afi. 1980. Biochim. Biophys. Acta **558**: 348–352.

A COMPARISON OF THE EFFECTS OF CHOLESTEROL AND 25-HYDROXYCHOLESTEROL ON EGG YOLK LECITHIN LIPOSOMES: SPIN LABEL STUDIES*

Gheorghe Benga,† Adriana Hodârnău,† Mihai Ionescu,‡ Victor I. Pop,† Petre T. Frangopol,‡ Vladimir Strujan,‡ Ross P. Holmes,§ and Fred A. Kummerow§

† *Department of Cell Biology*
Faculty of Medicine
Medical and Pharmaceutical Institute
Cluj-Napoca, Romania

‡ *Institute of Physics and Nuclear Engineering*
Bucharest-Măgurele, Romania

§ *Burnsides Research Laboratory*
Department of Food Science
University of Illinois at Urbana-Champaign
Urbana, Illinois 61801

Cholesterol is a widely distributed sterol found in many types of animal cell membranes.[1,2] It is believed to be an essential component of membranes, in particular the plasma membrane, which is its richest source. Much attention has focused on the interaction of cholesterol with membrane phospholipids and many biophysical techniques have been applied to the study of these interactions, including x-ray diffraction,[3,4] differential scanning calorimetry,[3,5] and a variety of spectroscopic techniques.[6] With these techniques, it has been shown that the fatty acyl chains of phospholipids, such as phosphatidylcholine (PC), which are in the disordered liquid-crystalline state, become less mobile because of hydrophobic interactions between cholesterol and PC (condensing effect of cholesterol). In contrast, the fatty acyl chains of PC that are in the ordered gel state become more mobile in the presence of cholesterol (liquefying effect of cholesterol). Thus, cholesterol modifies the motional freedom of fatty acyl chains in membranes, enhancing the movement of gel-state chains and decreasing that of fluid-state chains. Through studies mainly with liposomes, and to a lesser extent with monolayers, the parameters found to be central to this interaction with phospholipids are a 3β-hydroxyl group, a planar steroid-ring nucleus, and an isoprenoid side chain of precise size and saturation.[7,8]

Few studies have investigated the effects of low concentrations of cholesterol ($<$ 10 mol%) in membranes; the majority of studies have examined concentrations between 10–50 mol%. This is despite several biological membranes, including some types of endoplasmic reticulum, synaptic vesicles, and mitochondrial membranes, having cholesterol contents of less than 10 mol%.[1,8] The separation of lipids into different domains in membranes[9] suggests that cholesterol-poor domains could feasibly exist in cholesterol-rich membranes. Studies with filipin indicate that cholesterol-rich domains do indeed exist in membranes and that cholesterol is not homogeneously distributed in the plane of the membrane.[10,11] Thus, one objective of this study was to examine in detail the effects of cholesterol at low concentrations.

* Supported by the Academy of Medical Sciences and Ministry of Education and Teaching (Romania) and the National Science Foundation (USA).

0077-8923/83/0414–0140$01.75/0 © 1983, NYAS

The second objective was to study in parallel with the effects of cholesterol the ordering effects of low and high concentrations of 25-hydroxycholesterol on egg PC fatty acyl chains. 25-Hydroxycholesterol is one of the major auto-oxidation products of cholesterol.[12] It has been speculated that small amounts are synthesized in certain animal tissues *in vivo*.[13] Concern has been expressed over the possibility that substantial amounts of this sterol and other oxygenated derivatives are present in processed or deep-fried foods containing cholesterol.[14,15] Recent studies have shown that 25-hydroxycholesterol is an extremely potent atherogenic factor and that low concentrations are toxic to cultured, aortic smooth muscle cells.[16] Biochemical studies indicate that 25-hydroxycholesterol is an effective inhibitor of cholesterol biosynthesis, acting through the inhibition of the activity of a key regulatory enzyme in the biosynthetic pathway, 3-hydroxy-3-methyl glutaryl coenzyme A (HMGCoA) reductase.[13] The mechanism by which 25-hydroxycholesterol suppresses activity of this enzyme is not known. The effect is not direct, as the activity of microsomal preparations is not affected by the addition of inhibitory sterols.[17] Oxidized derivatives of cholesterol exert pleiotropic effects on cellular metabolism, perturbing such functions as DNA synthesis[18] and membrane permeability.[19] Attempts have been made to rationalize this multitude of effects as being solely related to the inhibition of sterol biosynthesis and a resultant decrease in the cholesterol content of cellular membranes.[13,19] However, the discordant effects of certain oxygenated sterols on sterol and DNA synthesis,[20] as well as the failure to detect altered membrane permeabilities in cells whose membrane cholesterol content was altered by a different means,[21] suggest that different mechanisms must be invoked to explain the entire range of effects of oxidized sterols. One plausible mechanism to explain some of the effects of 25-hydroxycholesterol, particularly its effect on membrane permeability, is its insertion into the membrane. It has been proposed that oxygenated sterols affect red blood cell morphology by this means.[22]

Studies on the effects of 25-hydroxycholesterol on the physical properties of membrane lipids are limited. A recent spin-labeling study using 5-doxylstearic acid reported that concentrations of up to 15 mol% of the sterol increase membrane order to a greater extent than cholesterol, whereas concentrations above 15 mol% produce no further increase in membrane order, in contrast to the effect of cholesterol.[23]

We report here our study of the interactions of cholesterol and 25-hydroxycholesterol with egg yolk PC in liposomes. We have used electron spin resonance (ESR) to analyze the spectra of various spin-labeled probes. This technique, because of its high sensitivity, has proved to be useful for investigating the effects of cholesterol on molecular interactions in membranes.[23-38] Although the bulk of the spin label may itself contribute to molecular disorder, results obtained with spin-labeling techniques agree qualitatively with those obtained by other techniques, such as NMR and fluorescence spectroscopy.

Experimental Procedures

Materials

Chromatographically pure egg yolk phosphatidylcholine was purchased from Lipid Products (South Nutfield, U.K.) or was prepared according to Singleton *et al.*[39] Cholesterol was purchased from Koch-Light (Coinbrook, U.K.) and 25-hydroxycholesterol from Steraloids (Wilton, NH). The spin labels (5-, 12-, or 16-doxylstearic acid) were obtained from Syva Corp. (Palo Alto, CA) and

TEMPO was synthesized according to Rozantsev.[40] All chemicals were analytical grade reagents.

Preparation and Labeling of Liposomes

Liposomes labeled with fatty acid spin probes were prepared by mixing egg PC dissolved in chloroform-methanol (2:1, vol/vol) sterol in chloroform-methanol (1:3, vol/vol) and the spin label ($\leqslant$ 1 mol%). Solvent was removed under nitrogen and then under vacuum for at least 2 hr. Water was added to hydrate the lipids and multilamellar dispersions (liposomes) were obtained by vigorous agitation (15–20 min). For obtaining liposomes labeled with TEMPO, the dried mixtures of egg PC and sterols were dispersed in a 5×10^{-3} M solution of TEMPO.

Electron Spin Resonance Measurements

The samples were sealed in glass capillaries and supported vertically in a narrow tube. The temperature was controlled by placing the tube in a glass/quartz Dewar flask through which a stream of preheated N_2 gas was passed. An ART-6 ESR spectrometer (manufactured by the Institute of Physics and Nuclear Engineering, Bucharest-Magurele, Romania) equipped with a variable temperature-control apparatus was used for all measurements. The sample temperature was monitored by a digital read-out device connected to a copper constantan thermocouple placed immediately below the sample. Where it is not stated otherwise, the temperature was maintained at 20°C. Instrumental parameters were optimized to avoid artifactual broadening.

Data presented here are means of two to four samples run under each condition. Order-parameter values (*S*), which are a measure of the distribution of molecular orientations relative to an axial plane perpendicular to the bilayer and usually correlated with the degree of membrane fluidity,[41] were calculated from the distances between the outer ($2T'_{\parallel}$) and inner ($2T'_{\perp}$) hyperfine spectral splittings (FIGURE 3), according to the method of Hubbell and McConnell,[25] with correction of the $T_{\perp}$ value by the method of Gaffney.[42] When the outer peaks were not well resolved (for example with the 12-doxyl spin label), we calculated the order parameter using the formula[43,44]:

$$S = \frac{43.7\ G - 3\ T_{\perp}'}{46.1} \times 1.723.$$

RESULTS

The small spin-label molecule TEMPO partitions between aqueous and hydrophobic regions in model membranes much as it does between hexane and water.[45] FIGURE 1 illustrates the spectrum of TEMPO in egg PC liposomes. Each such spectrum is a superposition of two spectra: one is the result of TEMPO dissolved in the fluid, hydrophobic region of the lipid and the other is the result of TEMPO in the aqueous phase. The high-field hyperfine line is split, the inner signal (with amplitude *H*) arising from TEMPO tumbling rapidly in the fluid lipid environment and the outer signal (with amplitude *P*) arising from TEMPO in water.[46] The TEMPO spectral parameter f, equal to $H/(H+P)$, which is correlated with the fraction of spin label dissolved in the membrane bilayer, thus gives an estimate of the solubility of the

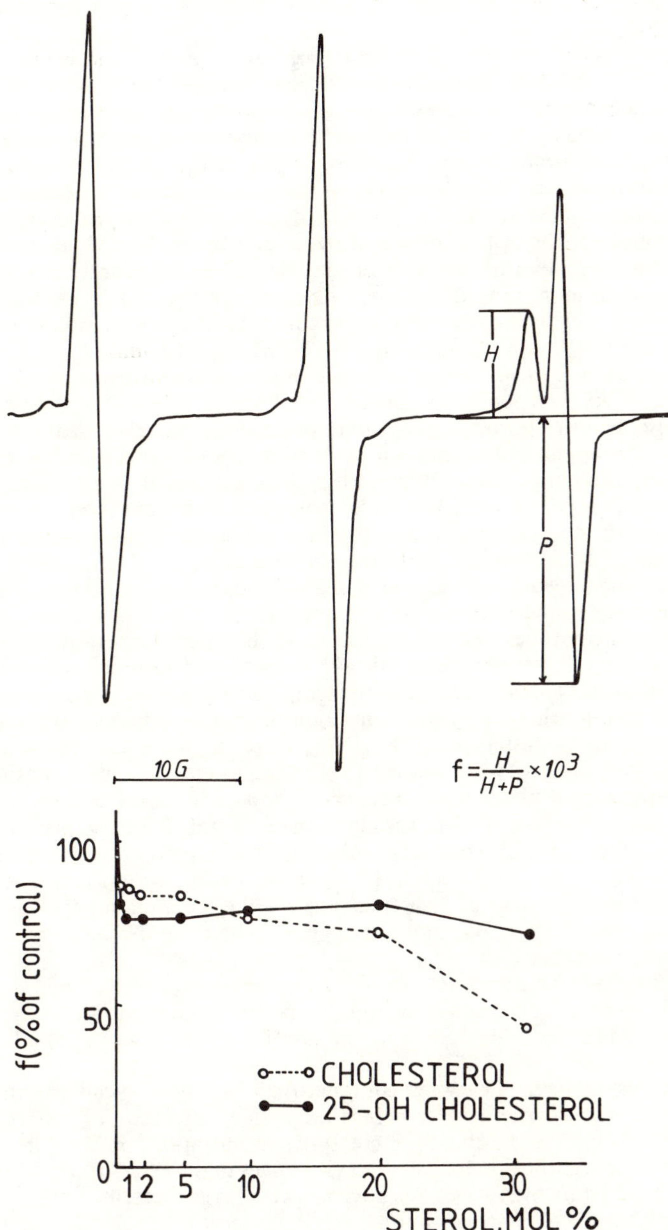

FIGURE 1. A typical ESR spectrum of TEMPO in egg PC liposomes and the variation in the spectral parameter (f) as a function of sterol concentration.

spin label in the membrane. Since the solubility of TEMPO in membranes depends on the fluidity of the lipid bilayer, the spectral parameter *f* provides an estimate of membrane fluidity.

The relationship between this spectral parameter and the concentration of cholesterol and 25-hydroxycholesterol in egg PC liposomes is shown in FIGURE 1. With increasing concentrations of cholesterol, a gradual decrease in the TEMPO parameter occurs. By contrast, 25-hydroxycholesterol produces a sharp decrease in *f* at very low concentrations (up to 1%), while above 2% no further decrease in the TEMPO parameter was observed. This shows (1) that cholesterol decreases the motional freedom of fatty acyl chains in egg PC liposomes in a concentration-dependent manner, a result in agreement with that obtained by other workers[26-29]; (2) that very small concentrations of 25-hydroxycholesterol may decrease significantly the fluidity of lipid bilayer, even more than the corresponding concentrations of cholesterol; and (3) that above 1 mol%, increasing concentrations of 25-hydroxycholesterol do not decrease membrane fluidity as measured by TEMPO partitioning.

Spin-labeled fatty acids are able to give more precise information on the lipid environment at known depths in the bilayer. FIGURE 2 shows typical spectra of 5-doxylstearic acid in liposomes containing cholesterol and 25-hydroxycholesterol, respectively. It is apparent that cholesterol at 50 mol% influences the spectrum considerably more than at 1 mol%. By contrast, the spectrum of 5-doxylstearic acid in liposomes containing 50 mol% 25-hydroxycholesterol is similar to the one containing 2 mol% sterol. It is also apparent that 25-hydroxycholesterol has a smaller influence on the spectra of 5-doxylstearic acid than cholesterol.

The motional freedom of fatty acyl chains in lipid bilayers is very sensitive to temperature changes and offers another experimental approach to probe lipid-lipid interactions in membranes. The interrelationship between temperature, sterol concentration, and the order parameter of both 5-doxyl and 12-doxylstearic acids was examined (FIGURES 3 and 4). Increasing the temperature decreased the order parameter with both spin-labeled fatty acids. The addition of sterols increased the order parameter. Similar to the results with TEMPO, differences in the effects of cholesterol and 25-hydroxycholesterol were observed. With cholesterol, a continuous increase in the order parameter occurred up to a concentration of 50 mol%. However, the rate of increase in order parameter was greater between 0 and 30 mol% than between 30 and 50 mol%. With 25-hydroxycholesterol and 5-doxylstearic acid as a probe, an increase in the order parameter occurred only in the concentration range of 0.1 to 5 mol% at all temperatures. Only a slight increase in the order parameter was detected at concentrations of 25-hydroxycholesterol greater than 5 mol% at 20°C and 37°C, and no increase at all at 10°C.

With 12-doxylstearic acid as a probe, an ordering effect of 25-hydroxycholesterol was observed at concentrations up to 1 mol%. Above this concentration, no further increase in order was observed. Similar effects were observed at all temperatures.

The greatest difference between cholesterol and 25-hydroxycholesterol was observed with 16-doxylstearic acid. Whereas increasing concentrations of cholesterol markedly altered spectra (FIGURE 5), increasing concentrations of 25-hydroxycholesterol produced no significant changes. These changes in the ESR spectra of 16-doxylstearic acid induced by the two sterols have been expressed in terms of the maximum hyperfine splitting, $2\ T'_{||}$ (*G*) in FIGURE 6. Cholesterol produced a concentration-dependent increase in $2\ T'_{||}$. The rate of change in this spectral parameter was much lower between 30 and 50 mol%, similar to the effects observed with 5-doxylstearic and 12-doxylstearic acids. At 10°C, 25-hydroxycholesterol had no effect. A slight effect was observed at 20°C with concentrations up to 2 mol%.

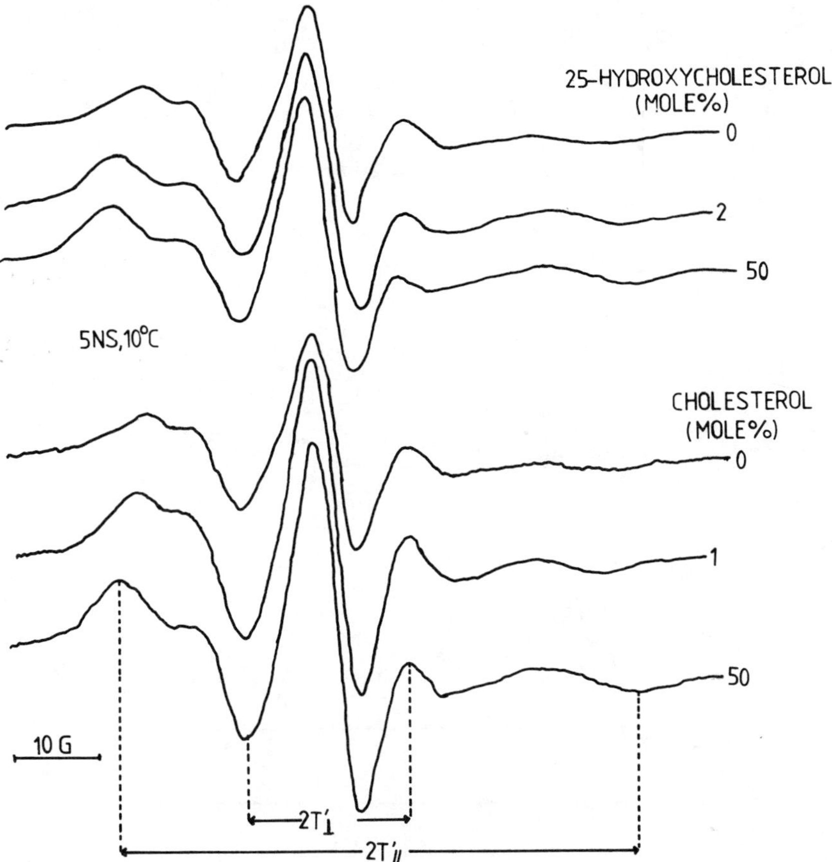

FIGURE 2. ESR spectra of 5-doxylstearic acid in egg PC liposomes containing varying concentrations of cholesterol and 25-hydroxycholesterol.

DISCUSSION

A pertinent question related to these and other spin-label studies on membranes is the reliability of the technique in providing information concerning the motional state of membrane lipids. Taylor and Smith[38] recently suggested that order parameters measured using doxyl spin-label probes do not faithfully represent the effects of external agents. These authors suggested that doxyl spin labels themselves produce large perturbations in membranes and give inaccurate responses to the addition of cholesterol. Their claims are based on measurements of the quadrupole splitting of deuterated stearic acid with and without an added 5-doxyl group in egg PC. The attachment of the doxyl group directly to the fatty acid between two deuterated carbon atoms reversed the usual ordering effect of cholesterol. However, they also reported that cholesterol increases ESR order parameters when the doxyl probe is used alone, in agreement with NMR measurements when only the deuterium probe is used. It is

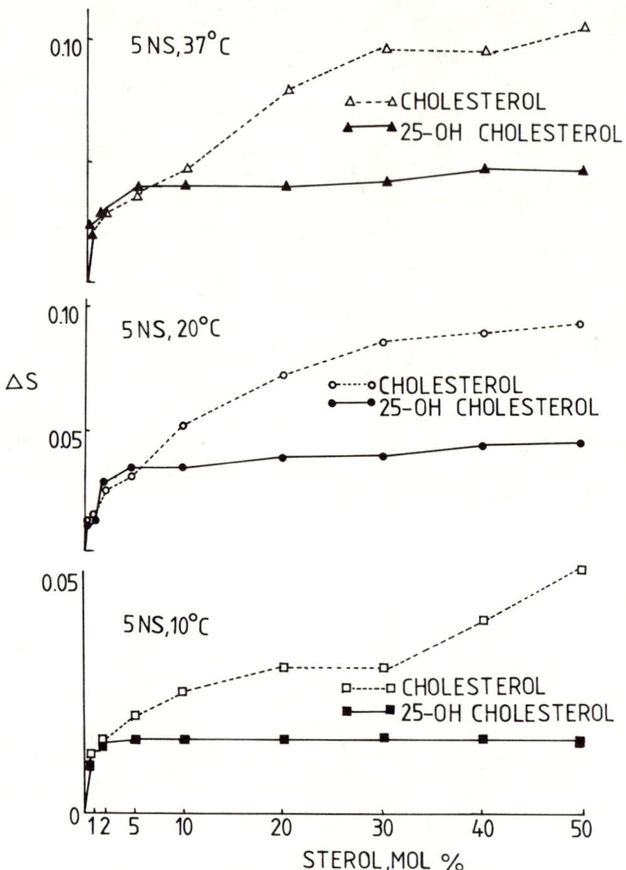

FIGURE 3. Change in order parameter (ΔS) of 5-doxylstearic acid–labeled liposomes as a function of sterol concentration at 10°C, 20°C, and 37°C,

reasonable to interpret their results as showing that the dual-labeled probe rather than the doxyl group perturbs the membrane.

Data obtained with any extrinsic probe must be viewed with caution since it is the property of the probe that is being directly measured and not that of the membrane lipids. While different estimates may be obtained quantitatively of the order parameter, S, when comparing ESR and NMR techniques, and while it is reasonable to propose the difference is caused by spin-label probe perturbations, qualitatively, the same trends are usually observed.[47] The available evidence does suggest that doxyl spin labels are sensitive and informative monitors of membrane organization and its modulation by external agents. Comparative studies, such as those reported here with cholesterol and 25-hydroxycholesterol, should be free of considerations of perturbations caused by the probe, as they should be similar in each instance if they occur.

An examination of the effect of small concentrations of cholesterol (0.1–5 mol%) on the motional freedom of egg yolk PC fatty acyl chains indicates that even

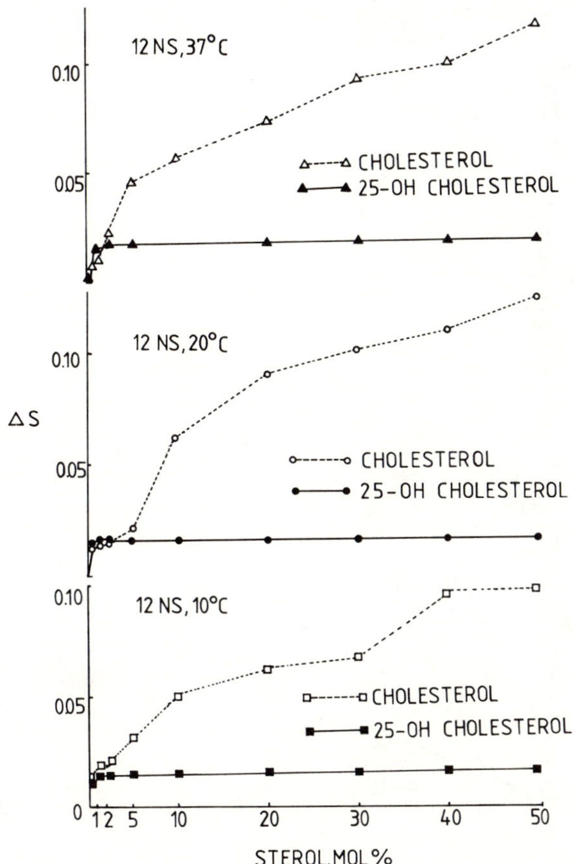

FIGURE 4. Change in order parameter (ΔS) of 12-doxylstearic acid–labeled liposomes as a function of sterol concentration at 10°C, 20°C, and 37°C.

at these concentrations cholesterol significantly affects the molecular order of membranes. This effect was observed to some extent with all spin-label probes and at all temperatures. The most pronounced effects were observed with 5-doxyl and 12-doxylstearic acids and only marginal effects with 16-doxylstearic acid at 20°C. These three spin-labeled fatty acids probe different planes within the lipid bilayer. The carboxyl head of the spin labels is positioned near the water interface and the hydrocarbon tail is generally parallel to the phospholipid hydrocarbon chains. In cholesterol-containing PC liposomes, 5-doxylstearic acid has its nitroxide moiety in close proximity to the phospholipid polar groups and close to the middle of the steroid nucleus of cholesterol.[36] The nitroxide group in 12-doxylstearic acid is located just below the steroid nucleus, whereas in 16-doxylstearic acid it lies deeper within the hydrocarbon interior of the membrane at a level just below the last methyl group of the isoprenoid tail.[36] Therefore, these results with low concentrations of cholesterol indicate that the sterol ring structure exerts the predominant effects in interactions with fatty acyl chains. Some studies have indicated that the permeability of membranes to

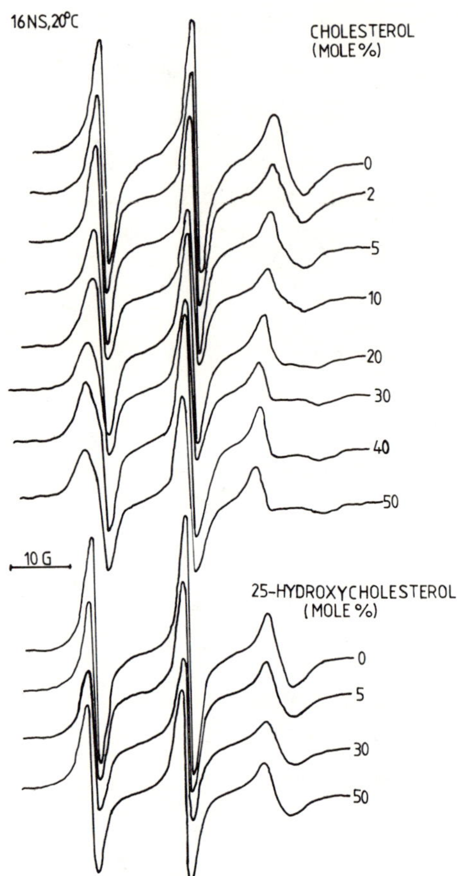

FIGURE 5. ESR spectra of 16-doxyl-stearic acid in egg PC liposomes containing varying concentrations of cholesterol and 25-hydroxylcholesterol.

water initially increases with concentrations of cholesterol to 10 mol%, but decreased with concentrations of cholesterol above 10 mol%.[48] These two divergent effects of low and high cholesterol concentrations are consistent with apparent changes in bilayer thickness that have been observed by x-ray diffraction in gel phase, dipalmitoylphosphatidylcholine-cholesterol mixtures where maximum thickness occurred at 7.5 mol% cholesterol,[3] and where presumably the tilt of the fatty acyl chains is straightened. Thus, measurements on water permeability, bilayer thickness, and the spectra of spin-label probes indicate that at low concentrations cholesterol has a significant effect on the properties of membrane phospholipids that differs from its effect at high concentrations.

The reasons for the effects of low concentrations of cholesterol may be related to the proposed packing, based on molecular models, of one cholesterol molecule with seven phospholipids.[49] Thus, if cholesterol-cholesterol interactions and acyl chain sharing do not occur at low cholesterol concentrations, at 14 mol% cholesterol complete association of all phospholipid molecules with a sterol would result. The pronounced effects occurring between 0 and 5 mol% cholesterol suggest that either more than one shell of phospholipid molecules are influenced by cholesterol or that

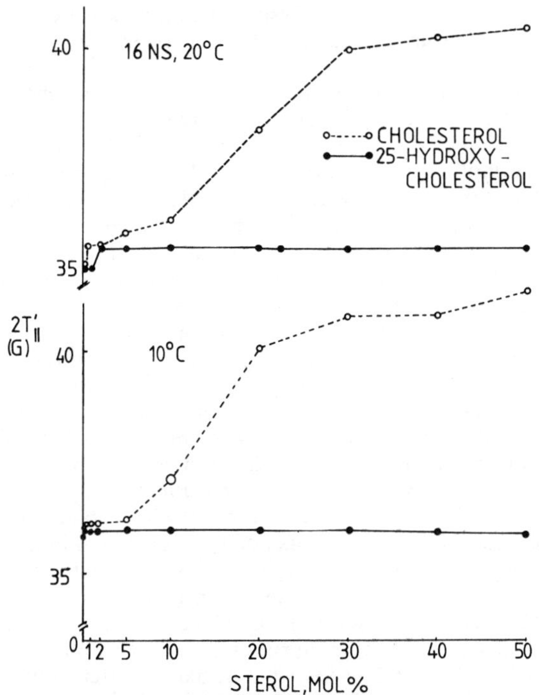

FIGURE 6. Change in the spectral parameter (2T') of 16-doxylstearic acid–labeled liposomes as a function of sterol concentration at 10°C and 20°C.

there is a preferential association of cholesterol with a particular species of PC in the egg yolk mixture. This response to concentrations of sterol up to 14 mol% may involve primarily the steroid nucleus. In studies where the length or bulk of the side chain has been modified, an ordering effect is observed up to 15 mol%.[36,50] At concentrations above this, either a change in the increase in order relative to that of cholesterol was observed, or in the case of ergosterol,[36] an actual decrease was observed. Another reason to consider is the effect of cholesterol at the bilayer surface. Neutron diffraction studies suggest that cholesterol does not directly interact with the phosphorylcholine headgroup.[51] The cholesterol hydroxyl group appears to be positioned near to the glycerol-fatty acyl ester linkages of the phospholipid. At low concentrations of cholesterol, phosphorylcholine dipole motion decreases, reaching a minimum between 20 and 30 mol%.[52] Motion increases above 30 mol%. That some change occurs in the relative orientation of charged moieties on the headgroup is evident from the cation selectivity of PC bilayer films at cholesterol concentrations up to 10 mol%, whereas anion selectivity appears in the 20–50 mol% range.[53]

The parameters derived from the use of TEMPO and the spin-labeled fatty acids as probes indicate that, as the cholesterol concentration is increased from 10 to 50 mol%, there is an increase in the molecular order of the phospholipid acyl chains. This occurred whether the parameter measured was the partitioning of TEMPO into the membrane, the order parameter with 5-doxyl and 12-doxylstearic acids, or the maximum hyperfine splitting with 16-doxylstearic acid. As expected, for a given lipid composition, a decrease in temperature also resulted in an increase in the ordering of membrane lipids (results not shown). Despite the increased ordering of acyl chains associated with a temperature decrease, cholesterol still exerted a substantial effect at all temperatures.

The increase in molecular order associated with increasing cholesterol is responsible for the condensing effect of cholesterol. Various models have been proposed to explain this condensing effect. Rothman and Engelman[54] attributed it to steric hindrance of the phospholipid hydrocarbon chain. A close approach by the sterol nucleus to the hydrocarbon chain restricts the number of conformations accessible to the upper portion of the chain. This results in reduced conformational freedom of the phospholipid chains, giving rise to an increase in order. Spin label[24,50,55] and other studies[7] have shown that specific structural requirements for cholesterol and its analogues are necessary before significant changes in the degree of order are observed. The sterol requirements are a planar steroid nucleus, a 3β-hydroxyl group, and a hydrocarbon chain at C-17 of precise length and saturation. This side chain is required to induce appreciable order in the bilayer.[24,50,55] Specifically, cholesterol analogues with side chains containing three or fewer carbons cause significantly less ordering.[50] NMR studies have indicated the side chain in cholesterol is highly flexible.[56] This may not be true of the entire side chain as it has been suggested that the methyl groups at C-13 and C-20 render that portion of the side chain inflexible.[57] The function of the cholesterol side-chain may possibly be to restrict the overall motion of the steroid nucleus, enhancing its ability to order the bilayer.[36]

It is also of significance that low concentrations of 25-hydroxycholesterol generally behaved similarly to cholesterol. Some differences were apparent at 5 mol%, in particular with 12-doxylstearic acid at 37°C. Concentrations of oxidized sterols in this range are obtained in erythrocyte membranes following the incubation of erythrocytes with 10 μg sterol/ml.[22] This insertion of oxidized sterols in the membrane is accompanied by morphological changes[22] and a diminished osmotic fragility.[58] It is clear from the spin-label studies in this report that perturbations in membrane structure related to the insertion of oxidized sterols at these levels cannot be detected using this technique, if indeed they do occur. At concentrations above 5 mol%, it is evident that 25-hydroxycholesterol differs markedly from cholesterol in not having the ability to increase molecular order with increasing concentrations. Thus, it appears that the hydroxyl group in the 25-position limits the interactions of the steroid nucleus with phospholipid acyl chains. This is further evidence for the importance of the precise structural requirements for the isoprenoid side chain in the effect of sterols on membranes. These results are in agreement with another report[24] that the presence of a polar group in the region of the hydrocarbon tail renders the steroid totally ineffective in increasing molecular order in membranes as demonstrated with pregnanalone, 5-pregnen-3β-ol-20-one, testosterone, cholic acid, and cortisone.

It is apparent from these results that cholesterol and 25-hydroxycholesterol exert different effects on membranes. Whether these differences can, in part, be responsible for the observed effects of oxidized sterols on the metabolism and permeability of cultured cells, the necrosis of aortic smooth muscle cells, and the development of atherosclerotic plaques remains to be clarified by further research.

Acknowledgments

The authors wish to thank Dr. L. Berliner (Ohio State University, Columbus, OH) for helpful discussions.

REFERENCES

1. NES, W. R. 1974. Lipids **9**: 596–612.
2. JAIN, M. K. 1975. Curr. Top. Membr. Trans. **6**: 1–57.
3. LADBROOKE, B. D., R. M. WILLIAMS & D. CHAPMAN. 1968. Biochim. Biophys. Acta **150**: 333–340.
4. LECUYER, H. & D. G. DERVICHIAN. 1969. J. Mol. Biol. **45**: 39–57.
5. DE KRUYFF, B., R. A. DEMEL, A. J. SLOTBOOM, L. L. M. VAN DEENEN & A. F. ROSENTHAL. 1973. Biochim. Biophys. Acta **307**: 1–19.
6. DEMEL, R. A. & B. DE KRUYFF. 1976. Biochim. Biophys. Acta **457**: 109–132.
7. DEMEL, R. A., K. R. BRUCKDORFER & L. L. M. VAN DEENEN. 1972. Biochim. Biophys. Acta **255**: 311–320.
8. GREEN, C. 1977. Int. Rev. Biochem. **14**: 101–152.
9. KARNOVSKY, M. J., A. M. KLEINFELD, R. L. HOOVER & R. D. KLAUSNER. 1982. J. Cell Biol. **94**: 1–6.
10. MONTESANO, R. 1979. Nature **280**: 328–329.
11. MONTESANO, R., A. PERRELET, P. VASSALLI & L. ORCI. 1979. Proc. Natl. Acad. Sci. USA **76**: 6391–6395.
12. SMITH, L. I. 1981. Cholesterol Autoxidation. Plenum Press. New York.
13. KANDUTSCH, A. A., H. W. CHEN & H. -J. HEINIGER. 1978. Science **201**: 498–501.
14. TAYLOR, C. B., S. -K. PENG, N. T. WERTHESSEN, P. THAM & K. T. LEE. 1979. Am. J. Clin. Nutr. **32**: 40–57.
15. KUMMEROW, F. A. 1979. Am J. Clin. Nutr. **32**: 58–83
16. PENG, S. -K., P. THAM, C. B. TAYLOR & B. MIKKELSON. 1979. Am. J. Clin. Nutr. **32**: 1033–1042.
17. KANDUTSCH, A. A. & H. W. CHEN. 1973. J. Biol. Chem. **248**: 8408–8417.
18. CHEN, H. W., A. A. KANDUTSCH & C. WAYMOUTH. 1974. Nature **251**: 419–421.
19. CHEN, H. W., H. -J. HEINIGER & A. A. KANDUTSCH. 1978. J. Biol. Chem. **253**: 3180–3185.
20. DEFAY, R., M. E. ASTRUC, S. ROUSSILLON, B. DESCOMPS & A. CRASTES DE PAULET. 1982. Biochem. Biophys. Res. Commun. **106**: 362–372.
21. BAKKER-GRUNWALD, T. & M. SINENSKY. 1979. Biochim. Biophys. Acta **558**: 296–306.
22. HSU, R. C., J. R. KANOFSKY & S. YACHNIN. 1980. Blood **56**: 109–117.
23. SINENSKY, M. 1981. Arch. Biochem. Biophys. **209**: 321–324.
24. BUTLER, K. W., I. C. P. SMITH & H. SCHNEIDER. 1970. Biochim. Biophys. Acta **219**: 514–517.
25. HUBBELL, W. L. & H. M. MCCONNELL. 1971. J. Am. Chem. Soc. **93**: 314–326.
26. HSIA, J. C., H. SCHNEIDER, & I. C. P. SMITH. 1970. Chem. Phys. Lipids **4**: 238–242.
27. HSIA, J. C., H. SCHNEIDER & I. C. P. SMITH. 1971. Can. J. Biochem. **49**: 614–622.
28. OLDFIELD, E. & D. CHAPMAN. 1972. FEBS Lett. **23**:285–297.
29. MARSH, D. & I. C. P. SMITH. 1973. Biochim. Biophys. Acta **298**: 133–144.
30. SCHREIER-MUCCILLO, S., K. W. BUTLER & I. C. P. SMITH. 1973. Arch. Biochem. Biophys. **159**: 297–311.
31. SHIMSHICK, E. J. & H. M. MCCONNELL. 1973. Biochem. Biophys. Res. Commun. **53**: 446–451.
32. MAILER, C., C. P. S. TAYLOR, S. SCHREIER-MUCCILLO & I. C. P. SMITH. 1974. Arch. Biochem. Biophys. **163**: 671–678.
33. SINGER, M. A. & J. K. S. WAN. 1975. Can. J. Physiol. Pharmacol. **53**: 1065–1071.
34. BUTLER, K. W. & I. C. P. SMITH. 1978. Can. J. Biochem. **56**: 117–122.
35. PANG, K. Y. Y. & K. W. MILLER. 1978. Biochim. Biophys. Acta **511**: 1–9.
36. SEMER, R. & E. GELERINTER. 1979. Chem. Phys. Lipids **23**: 201–211.
37. CHIN, J. H. & D. B. GOLDSTEIN. 1981. Mol. Pharmacol. **19**: 425–431.
38. TAYLOR, M. G. & I. C. P. SMITH. 1980. Biochim. Biophys. Acta **599**: 140–149.
39. SINGLETON, W. S., M. S. GRAY, M. L. BROWN & J. L. WHITE. 1965. J. Am. Oil Chem. Soc. **42**: 53–56.
40. ROZANTSEV, E. G. 1970. Free Nitroxyl Radicals. Plenum Press. New York.
41. SCHREIER, S., C. F. POLNASZEK & I. C. P. SMITH. 1978. Biochim. Biophys. Acta **515**: 395–436.

42. Gaffney, B. J. 1976. *In* Spin Labeling. Theory and Applications. pp. 567–571. Academic Press. New York.
43. Bales, B., E. Lesin & S. Oppenheimer. 1977. Biochim. Biophys. Acta **465**: 400–407.
44. King, M. E., B. W. Stavens & A. A. Spector. 1977. Biochemistry **16**: 5280–5285.
45. Hubbell, W. L. & H. M. McConnell. 1968. Proc. Natl. Acad. Sci. USA **61**: 12–16.
46. Shimshick E. J. & H. M. McConnell. 1973. Biochemistry **12**: 2351–2360.
47. Stockton, G. W., C. F. Polanszek, A. P. Tulloch, F. Hasan & I. C. P. Smith. 1976. Biochemistry **15**: 954–966.
48. Jain, M. K., D. G. Touissaint & E. H. Cordes. 1973. J. Membr. Biol. **14**: 1–16.
49. Engelman, D. M. & J. E. Rothman. 1972. J. Biol. Chem. **247**: 3694–3697.
50. Suckling, K. E. & G. S. Boyd. 1976. Biochim. Biophys. Acta **436**: 295–300.
51. Worcester, D. L. & N. P. Franks. 1976. J. Molec. Biol. **100**: 359–378.
52. Shepherd, J. C. W. & G. Buldt. 1979. Biochim. Biophys. Acta **558**: 41–47.
53. Scibona, G., B. Scuppa, C. Fabiani & M. Pizzichini. 1978. Biochim. Biophys. Acta **512**: 41–53.
54. Rothman, J. E. & D. M. Engelman. 1972. Nature New Biol. **237**: 42–44.
55. Hsia, J. C., R. A. Long, F. E. Hruska & H. D. Gesser. 1972. Biochim. Biophys. Acta **290**: 22–31.
56. Kroon, P. A., M. Kainosho & S. I. Chan. 1975. Nature **256**: 582–584.
57. Oldfield, E., M. Meadows, D. Rice & R. Jacobs. 1978. Biochemistry **17**: 2727–2740.
58. Streuli, R. A., J. R. Kanofsky, R. B. Gunn & S. Yachnin. 1981. Blood **58**: 317–325.

THE SPIN-LABEL APPROACH TO LABELING MEMBRANE PROTEIN SULFHYDRYL GROUPS*

Lawrence J. Berliner

Department of Chemistry
Ohio State University
Columbus, Ohio 43210

The membrane of most eukaryotic cells comprises lipid, cholesterol, and protein molecules. The precise three-dimensional arrangement of the components of a cell membrane imparts its unique functions via several specific molecular interactions. The fluid-mosaic model of Singer and Nicholson gives us a picture of two asymmetric lipid lamellae into which proteins are intercalated to varying extents.[1] The lipid phase of the membrane is also arranged asymmetrically in the familiar bilayer structure. Basically, the two different types of proteins are classified by Singer and Nicholson,[1] integral and peripheral, depending on their ease of isolation from the membrane. Peripheral proteins are generally at the bilayer surface, held (loosely) by electrostatic interactions; integral proteins penetrate the lipid phase of the membrane to varying degrees and frequently traverse the entire bilayer.

A particularly well-studied membrane system, the erythrocyte, exemplifies several of the aspects of membrane structure discussed above. Its phospholipids are asymmetrically distributed. The principal peripheral cytoplasmic proteins have been characterized (spectrin, actin, band 4.1,G-3-PDH, etc.).[2] The chief integral proteins are band 3 and glycophorin.[3] There has been great interest and activity in past years in spin-labeling membrane proteins, particularly those of the red blood cell, in order to learn more about the relationships between protein and lipid structure, membrane function, disease, aging, etc. (for a recent review see, for example, Butterfield[4]). The spin-label technique has been instrumental in providing molecular information about the conformation of proteins, structure and function of membranes, nucleic acids, and other biological systems (for a complete background, see Berliner[5,6]). The usefulness of the nitroxide electron spin resonance (ESR) spin labels is in part their extreme sensitivity, their efficacy with opaque as well as optically clear samples, the molecular information about the local environment near the spin label, the chemical stability of nitroxides, and the relative simplicity of labeling and measurement. It is this last aspect, the relative ease and simplicity that occasionally obscures the validity of an experiment that employs a poorly chosen spin label. While the approach of using lipid-type spin labels to probe the phospholipid environment has been one method of attack in membranes studies, the other involves covalently labeling (specific) membrane proteins with nitroxide protein-modification agents, particularly those directed towards cysteine-thiol groups. This report reviews the specificity and efficacy of these labels, their advantages and pitfalls, and introduces a new spin label that shows great promise for future membrane and enzymological studies using this method.

* Supported by grants from The National Science Foundation (PCM77-24658) and the National Institutes of Health (HL 24549).

0077-8923/83/0414-0153$01.75/0

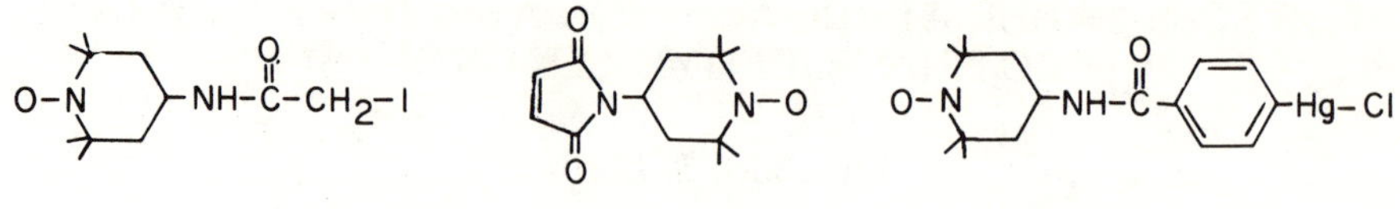

I II III

FIGURE 1. Structure of some sulfhydryl spin labels.

SPIN-LABEL THIOL MODIFICATION REAGENTS

The desirable properties of an excellent sulfhydryl spin label for membrane proteins are: (1) react rapidly and with absolute specificity; (2) contain the least number of flexible single bonds of rotation between the sulfhydryl group and the nitroxide ring in order to reflect most sensitively the conformation at the thiol group site; and (3) be susceptible to no other side reactions or decomposition.

FIGURE 1 depicts the structures of the types of (covalent) spin-label "sulfhydryl reagents" that have been used most frequently in the past. We first review their general specificity, examine one or two specific applications, and summarize their general promise as membrane-protein–directed reagents. Structure **I**, an iodoacetamide (piperidine) spin label, has been used fairly extensively with enzymes to label specific sulfhydryl groups (see, for example, the chapter of Morrisett in Berliner.[5] This label is also found as a pyrrolidine iodoacetamide, from the corresponding amine pyrrolidine precursor, but suffers slightly from the fact that the pyrrolidine nitroxides are a racemic mixture of two stereoisomers. These labels are, of course, analogues of haloacetates, (e.g., iodoacetate), which undergo the same general type(s) of reactions. Scheme I summarizes the type(s) of modifications expected between one of these labels and a protein, where Ⓟ signifies the protein macromolecule:

HALOACETATES

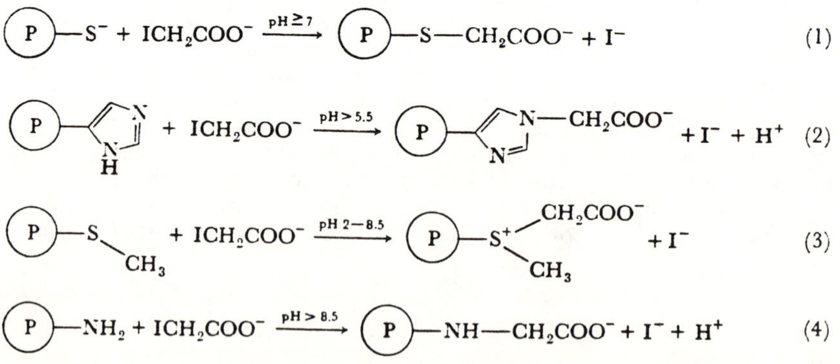

$$\text{Ⓟ}\!-\!S^- + ICH_2COO^- \xrightarrow{pH \geq 7} \text{Ⓟ}\!-\!S\!-\!CH_2COO^- + I^- \quad (1)$$

$$\text{Ⓟ}\!-\!\text{(imidazole, N-H)} + ICH_2COO^- \xrightarrow{pH > 5.5} \text{Ⓟ}\!-\!\text{(imidazole)}\!-\!N\!-\!CH_2COO^- + I^- + H^+ \quad (2)$$

$$\text{Ⓟ}\!-\!S\!-\!CH_3 + ICH_2COO^- \xrightarrow{pH\ 2-8.5} \text{Ⓟ}\!-\!S^+(CH_2COO^-)(CH_3) + I^- \quad (3)$$

$$\text{Ⓟ}\!-\!NH_2 + ICH_2COO^- \xrightarrow{pH > 8.5} \text{Ⓟ}\!-\!NH\!-\!CH_2COO^- + I^- + H^+ \quad (4)$$

While haloacetates are usually most reactive with thiol groups (reaction 1), under conditions of a many fold excess of reagent and long reaction times, at least one (or two) of the other types of moieties shown in Scheme I will have been partially modified, leading to a mixture of labeled groups and a nonintegral stoichiometry of labeling. With regard to spin-label experiments, a multicomponent ESR spectrum is frequently obtained, denoting several conformational environments. The promise for this reagent as a thiol reagent is obviously questionable when relatively harsh conditions (50–100-fold excess, several hours of labeling) are employed. Under very harsh conditions, even phenolic side chains of tyrosine have been found to react with haloacetates.[7] For these reasons this label type (**I**) has not been utilized very often in more recent years and less frequently in membrane-protein–labeling work, in contrast to the maleimide piperidine nitroxides.

Structure **II**, the maleimide piperidine nitroxide, is perhaps the most ubiquitous covalent spin label for protein-labeling work. It is fair to estimate that over 75% of all spin-labeling papers utilizing covalent labels have employed either that maleimide label **II** in FIGURE 1 or its pyrrolidinyl analogue(s). Scheme II below summarizes the most common reactions of maleimide groups with proteins:

MALEIMIDES

$$\text{(P)}\!-\!SH + \text{maleimide-}NC_2H_5 \xrightarrow{pH>5} \text{(P)}\!-\!S\!-\!\text{succinimide-}NC_2H_5 \qquad (1)$$

$$\text{(P)}\!-\!NH_2 + \text{maleimide-}NC_2H_5 \xrightarrow{pH>7} \text{(P)}\!-\!NH\!-\!\text{succinimide-}NC_2H_5 \qquad (2)$$

Since mercaptide ion is the most reactive species, there is marked enhancement with increasing pH. At pH 7 thiol reactivity can be about 1,000-fold greater than that with amino groups; however, at higher pH the reaction with lysine, histidine, proline, and other amines becomes more pronounced.[7] Again here, where the reagent is in many fold excess for long reaction times, the probability of partial amino group labeling is significant.

FIGURE 2 depicts a typical ESR spectrum of the piperidine maleimide spin label **II** attached to membrane proteins of the erythrocyte. The two general classes of spectral components, S and W, denote strongly and weakly immobilized nitroxide groups in two general classes of conformational environments. The narrow higher peak height component, W, which is actually a much smaller component of the total spin-label population (on a molar basis), is nonetheless altered very easily by small changes in the amount of label at W-type sites, since the peak height changes in direct proportion to the concentration of label in that particular conformational state. As is exemplified in Butterfield[4] as well as other studies of maleimide-labeled cells and membranes, the ratio of the W to S component is frequently the only parameter that may be used to correlate membrane state and function. Not only is there the general ambiguity in such complex systems about the specific proteins labeled, but also whether (highly mobile, surface exposed) lysine amino groups are modified as well. Typically, a high molar ratio of this label to protein is employed for periods of

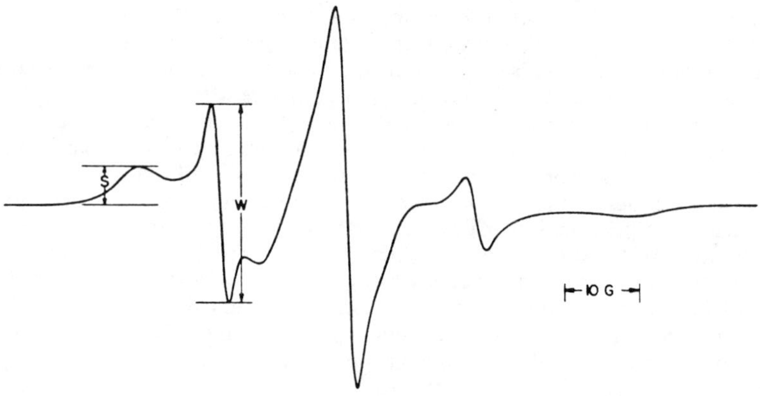

FIGURE 2. Typical ESR spectrum of the piperidine maleimide spin label (**II**) attached to membrane proteins in normal erythrocyte membranes. The amplitudes of the $M_I = +1$ lines of the spin label covalently bound to strongly and weakly immobilized binding sites are indicated by S and W, respectively. Fresh erythrocyte ghosts were labeled overnight at 4°C by spin label **II**. After five washes to remove excess spin label the spectra of the spin-labeled ghosts were recorded at room temperature with a sweep width of 100 G and modulation amplitude and microwave power incident on the resonant cavity of 0.32 G and 14 mW, respectively, using a Varian E-109 spectrometer. (From Butterfield.[4] With permission from Plenum Publishing Corp.).

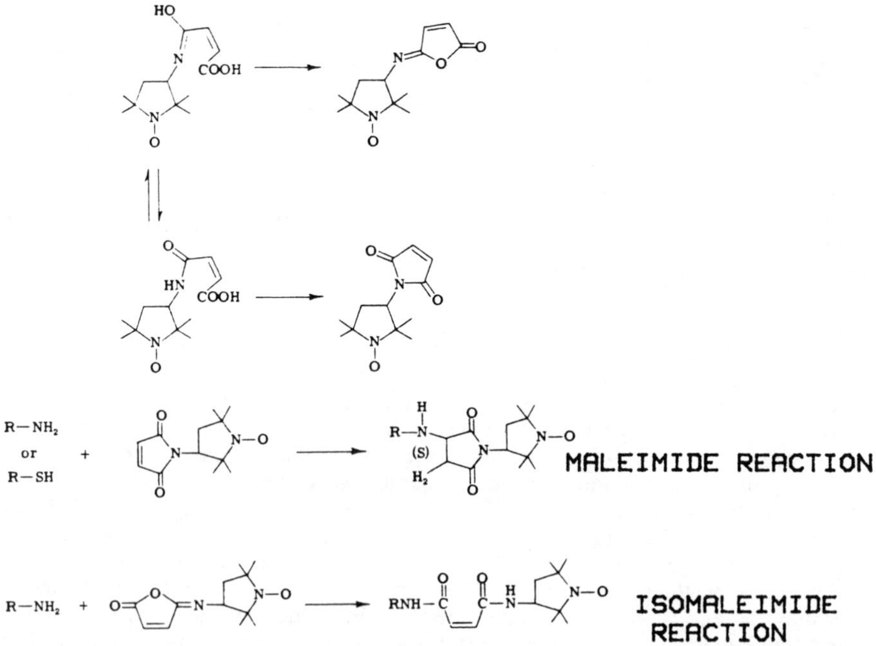

MALEIMIDE REACTION

ISOMALEIMIDE REACTION

FIGURE 3. Maleimide and isomaleimide syntheses. (From Berliner.[5] With permission from Academic Press, Inc.).

several hours to days.[4] Another problem with maleimides, which has been largely overcome in recent years, involves the synthetic scheme for their preparation. FIGURE 3 depicts the two possible products obtained in the "ring closing" step in maleimide synthesis—the desired maleimide and the isomaleimide. Upon reaction with a protein thiol, each yields a unique product, each of which would not be expected to give the same mobility in an ESR spectrum, by virtue of the differences in the flexible "links" between the nitroxide ring and the thiol group.[8] If one uses some care by thoroughly checking the purity of the maleimide spin label by thin layer chromatography and other methods, assurance of the maleimide **II** only is satisfied. While it was mentioned above that label **II** is the most popular protein spin label in use, its promise as a good thiol reagent must be presented with some reservations; most significantly, the broader specificity to protein side chains.

While the two classes of labels described above are irreversible modification reagents, it is frequently desirable to be able to reactivate a modified protein for further (control) studies. Structure **III** falls into the general class of mercurials, which react specifically and rapidly with sulfhydryl groups of proteins. While this type of label meets most of the requirements of a good thiol spin label dictated earlier, it suffers from an unsolved photodecomposition problem upon short exposures to normal room lighting as well as the relatively long distance between the nitroxide and thiol group.[9]

A New Optimal, Reversible Thiol-specific Spin Label

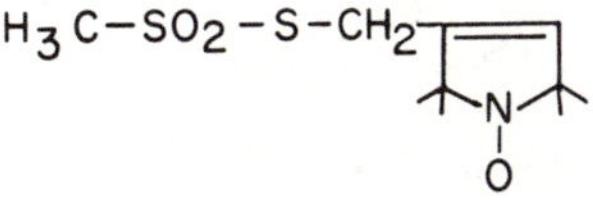

$H_3C-SO_2-S-CH_2-$ N–O

IV

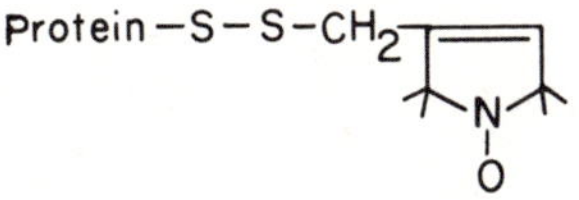

Protein–S–S–CH_2– N–O

V

The label **IV** (1-oxyl-2,2,5,5-tetramethylpyrroline-3-methyl)methanethiolsulfonate, whose synthesis is shown in FIGURE 4, is based on the elegant work of Kenyon[10] with alkane-thiosulfonates as reversible thiol reagents of selected bulkiness and hydrophobicity. As a model for a thiol-containing protein that can be specifically labeled and monitored by several techniques, we chose the enzyme *Papaya latex* papain, which contains a single active cysteine in the reduced form as the catalytic group of the active enzyme. If this SH group is modified, the enzyme loses all activity. The enzyme was inhibited rapidly and specifically with label **IV** within five min-

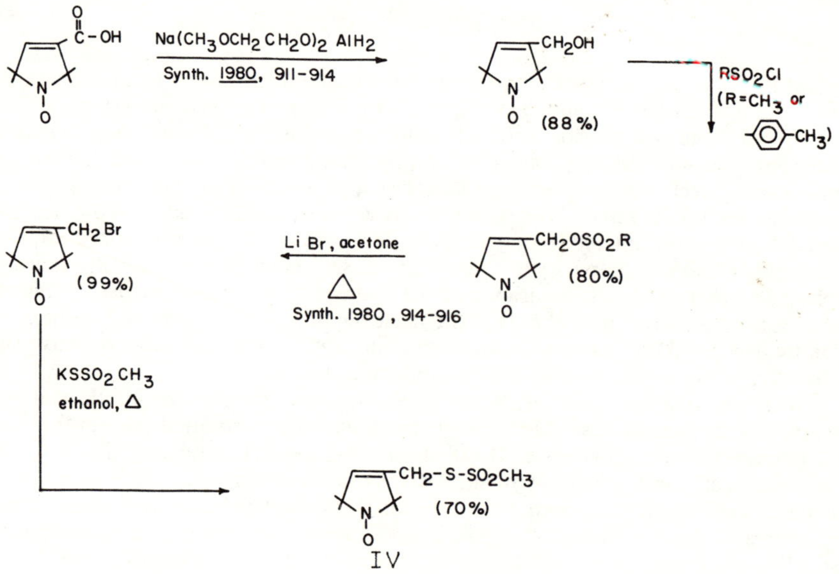

$Na(CH_3OCH_2CH_2O)_2AlH_2$

Synth. 1980, 911-914

CH_2OH (88%)

RSO_2Cl ($R=CH_3$ or $C_6H_4CH_3$)

CH_2OSO_2R (80%)

Li Br, acetone, Δ

Synth. 1980, 914-916

CH_2Br (99%)

$KSSO_2CH_3$ ethanol, Δ

CH_2-S-SO_2CH_3 (70%)

IV

FIGURE 4. Synthesis of the new reversible thiol spin label **IV**.

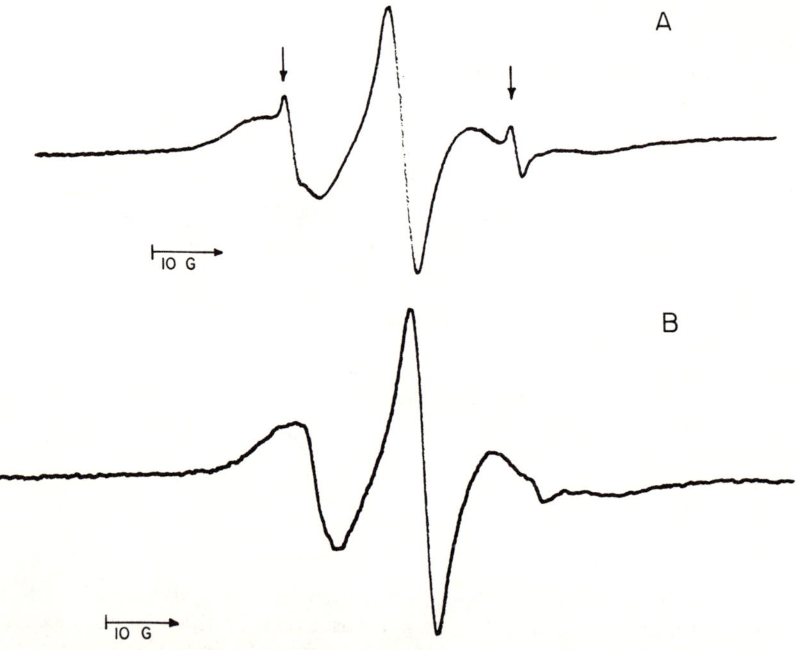

FIGURE 5. X-Band ESR spectra of active site spin-labeled papain. (A) Reaction mixture (pH 4.5) after 5 min of labeling where [papain] = 0.19 mM active sites and [label **IV**] = 0.1 mM. Buffer was 0.05 M acetate. The exceedingly small narrow line spectra (arrows) represents unreacted spin label, which accounts for *ca.* ± 0.5% of the total ESR signal.† (B) After reaction with excess spin label **IV** and elution from Sephadex G-10 at pH 7.5 (0.2 mM Tris-HCl, 0.025 mM EDTA). Papain (active site) concentration was 0.16 mM. (From Berliner *et al.*[11] With permission from *Analytical Biochemistry*.)

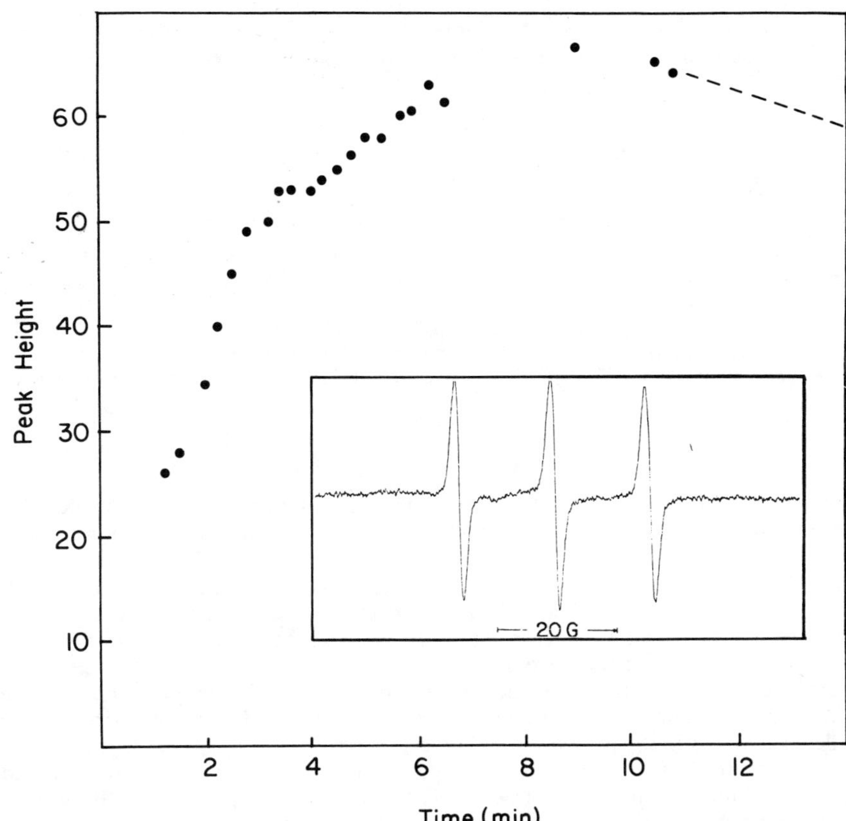

FIGURE 6. Spin-label release reaction monitored by ESR. Change of peak height of low-, center-, or high-field line of the growing "free" spin-label spectrum (see inset) with time. Conditions were: papain (0.098 mM active sites), DTT (1.7 mM), pH 7.5 (elution buffer), 26 ± 2°C. (From Berliner *et al.*[11] With permission from *Analytical Biochemistry*.)

utes as evidenced by a complete loss of esterase activity where label was in a slight excess. In FIGURE 5 the ESR spectra were obtained either directly from a reaction mixture where [papain] ⩾ [spin label] (FIGURE 5A)† or from a labeled protein purified on Sephadex G-10 (FIGURE 5B). (The spectra were identical whether measured at pH 4.5 or 7.5.) The high- and low-field hyperfine extrema were separated by 46.5 ± 1.0 G (intermediate immobilization), which is a very sensitive range to protein conformational changes. The specificity of this label was confirmed by reducing the disulfide linkage (**V**, FIGURE 4) between the active Cys 25 and the 3-methylpyrroline nitroxide with 1.0 to 10 mM dithiothreitol (DTT) as followed by the return of enzymatic activity or the increase in concentration of free spin label. The esterase activity was followed in the pH stat. Upon exposure of 0.16 mM spin-labeled enzyme to 1.7 mM DTT, the broad immobilized spectrum (e.g., FIGURE 5B) converted com-

† The very small residual free line component in the top spectrum (arrows) from a reaction using excess enzyme represents *ca.* 0.5% of the total spin label in the sample and testifies to both the purity and exceptional reactivity of this label.

TABLE 1

SUMMARY OF LABELING AND RELEASE (REACTIVATION) REACTION*

Protein (mM)	Initial DTNB Active Sites (mM)	Initial Activity† (U/mg)	Released Spin Label (mM)	Final Activity‡ (U/mg)	Active Sites/ Protein (%)	Spin Label/ Protein (%)
Control papain (0.60)	0.19	3.5 ± 0.5	-	8.5 ± 0.5	32 ± 3	–
Spin-labeled papain (0.098)	0.031	0	0.034§	8.9 ± 0.5	32 ± 3	35 ± 5

* Adapted from Berliner *et al.*[11]
† Measured in the absence of cysteine, DTT, or other reducing agents.
‡ Measured within two minutes after exposure to DTT (10 mM) or cysteine.
§ Measured in 1.7 mM DTT (pH 7.5) by ESR from the maximum peak height (*ca.* 10 minutes after mixing).

pletely to a narrow three-line "free" spectrum (FIGURE 6, inset). The released-label concentration corresponded directly to the enzyme active site (thiol) concentration. TABLE 1 depicts, for example, a typical experiment with a papain sample that contained 32 ± 3% active (thiol) sites by dithiobisnitrobenzoic acid (DTNB) and 35 ± 5% spin label **IV** incorporation after labeling. If the free spectrum were monitored for longer periods, a slow decrease in peak height was observed, due to the well-documented destruction of the nitroxide group by thiols such as DTT.‡ This new spin label **IV** is a highly reactive thiol-specific reagent with a minimum number of flexible linkages between the nitroxide ring and cysteine side chain. Equally important is the facile reversibility in the presence of relatively mild reducing agents, a distinct advantage where the protein or membrane under study is available in only very limited quantities. Since the liberated "free nitroxide" yields a sharp three-line ESR spectrum, the release reaction serves as a mechanism for assaying reactive thiol groups on a protein or membrane. It is estimated that the sensitivity of this method approaches concentrations in the 10^{-7} to 10^{-8} M range, depending on the signal-to-noise characteristics of the user's particular ESR spectrometer.

SUMMARY

Labeling experiments with viable membrane preparations are optimally undertaken where conditions are extremely mild (minimal excess of modification reagent, very short reaction times) in order to reduce the possibilities of introducing artifacts (cell aging and death, denaturation, etc.). Furthermore, specificity of labeling is a critical requirement where the system is complex and contains many proteins. The spin label **IV** described above offers several advantages over previously employed reagents. By virtue of its reversibility and ease of quantitation, one may repetitively label and remove label from the same membrane preparation in order to check reproducibility while using the same sample as a control.

‡We recommend carrying out the "release reaction" at DTT concentrations less than 2 mM for an accurate estimate of released spin-label concentration where the thiol nitroxide side reaction is slow by comparison (see chapter by B. G. Gaffney in Berliner[5]).

ACKNOWLEDGMENT

It is a pleasure to acknowledge my collaborators in this work, Drs. Jacob Grunwald, H. Olga Hankovszky, and Kalman Hideg.

REFERENCES

1. SINGER, S. J. & G. L. NICHOLSON. 1972. The fluid mosaic model of the structure of cell membranes. Science **175**: 720–731.
2. MARCHESI, V. T. 1979. Functional proteins of the human red blood-cell membrane. Sem. Hematol. **16**: 3–20.
3. STECK, T. L. & J. YU. 1973. Selective solubilization of proteins from red blood-cell membranes by protein perturbants. J. Supramolec. Struct. **1**: 220.
4. BUTTERFIELD, D. A. 1982. Spin labeling in disease. *In* Biological Magnetic Resonance. L. J. Berliner & J. Reuben, Eds. **4**: 1–78. Plenum Publ. Co. New York.
5. BERLINER, L. J., Ed. 1976. Spin Labeling: Theory and Applications. Academic Press. New York.
6. BERLINER, L. J., Ed. 1979. Spin Labeling: Theory and Applications II. Academic Press. New York.
7. MEANS, G. E. & R. E. FEENEY. 1971. Chemical Modification of Proteins. Holden-Day, Inc. San Francisco.
8. BARRATT, M. D., A. P. DAVIES & M. T. A. EVANS. 1971. Maleimide and isomaleimide pyrrolidine-nitroxide spin labels. Eur. J. Biochem. **24**: 280–283.
9. BOEYENS, J. C. A. & H. M. MCCONNELL. 1966. Spin-labeled hemoglobin. Proc. Natl. Acad. Sci. USA **56**: 22–25.
10. KENYON, G. A. & T. W. BRUICE. 1977. Novel sulfhydryl reagents. Methods Enzymol. **47**: 407–430.
11. BERLINER, L. J., J. GRUNWALD, H. O. HANKOVSZKY & K. HIDEG. 1982. A novel reversible thiol-specific spin label: papain active site labeling and inhibition. Anal. Biochem. **119**: 450–455.

ANALYSIS OF SPIN-LABELED ERYTHROCYTE MEMBRANES*

Leslie W.-M. Fung

Department of Chemistry
Loyola University of Chicago
Chicago, Illinois 60626

Some of the earlier electron paramagnetic resonance (EPR) studies of proteins labeled with *N*-(1-oxyl-2,2,6,6-tetramethyl-4-piperidinyl)maleimide (Mal-6) in intact erythrocyte membranes were mainly qualitative in nature and showed similar, but not identical results.[5,27] However, they demonstrated the potential of the EPR method in providing dynamic information for proteins in red cell membranes. Our biochemical understanding of the erythrocyte membrane components was also limited at that time. Following the electrophoretic identification of several distinct proteins in the membranes,[7] the basic knowledge of the chemical composition, molecular arrangement, and functional properties of erythrocyte membranes has flourished in recent years.[23,29] Meanwhile, new EPR techniques have been developed for studying molecules with slow motions.[30] We have compared two of the saturation transfer (ST) EPR techniques (V_2' and U_1') with the conventional EPR method (V_1) in monitoring molecules in erythrocyte membranes and have shown that V_2' detection is useful for these systems.[8] Thus, with the recent enhancements in both red cell membrane biochemistry and EPR techniques, the spin-labeled membrane protein system in erythrocytes can now be studied more quantitatively.

Biochemical Properties of the Spin Labels in Membranes

Mal-6 appears to alkylate mainly the sulfhydryl (SH) groups.[5,27] A very small amount of amino groups may also be labeled.[5,25] Our recent results show that the EPR signal of membranes depends strongly on the labeling method.[12] In order to compare results from different studies, spin-labeling methods should be specified. In our studies, Mal-6 is interacted with washed membrane ghosts at a weight ratio of about 30–50 μg per mg protein for one hour at pH 8, 4°C, in the dark. EPR double integrations of the V_1 spectra show that about 26.0 ± 2.9 nmoles of label are bound to one mg of protein.[14] Further calculation shows that there are about $8.3 \pm 0.7 \times 10^6$ labels per ghost. This is in good agreement with the earlier studies.[27] Assuming that the total SH content in membrane is about 130 nmoles per mg protein,[1] about 20% of the membrane SH groups are alkylated.

Spin-labeled right-side–out vesicles were prepared with and without ascorbic acid inside or outside the vesicle. Since ascorbic acid reduces the spin label, those labels that are in contact with ascorbic acid will become EPR silent. Results of double integration of signals show that about 80 ± 5% of the labels are on the cytoplasmic surface of the erythrocyte membrane and about 10 ± 2% on the extracellular surface.[14] About 10% of the signal is unaccounted for in these studies. Recently, we

* Supported in part by grants from Michigan Heart Association, Huntington Disease Foundation, and the National Institutes of Health (HL-22432 and HL-16008). L.W.-M.F. is a National Institutes of Health Research Career Development Awardee (KO4 HL-00860).

0077-8923/83/0414–0162$01.75/0 © 1983, NYAS

have isolated proteins from labeled, intact membranes by various extraction methods; preliminary results show about 20 labels per spectrin-actin complex, 1 label per Band III, and 10 labels on other proteins (L. Fung and W. Meena, unpublished results). In order to obtain these results, the values of number of copies of individual proteins in membranes provided by Steck[29] are used. Furthermore, our calculations show that, in the intact membrane, a large amount of the labels (over 50%) is associated with the spectrin-actin complex. This result is in accord with the results of an earlier radioactive *N*-ethyl maleimide labeling study.[4]

EPR Spectral Properties of Conventional EPR (V_1) *Detection and Their Applications: W/S Ratios*

The V_1 spectrum of the Mal-6–labeled membranes consists of two components, a narrow-line and a broad-line component.[5,27] Many workers have used the amplitude ratio, W/S (FIGURE 1), to analyze their spectra.[3,9,16,19–21,25,28] Most of the W/S ratios in these studies are qualitatively similar, but not quantitatively the same. Recent studies show that the W/S value is very sensitive to the physical state of the membrane. For example, experimental parameters, such as temperature, pH, ionic strength, etc., will cause large changes in the W/S value.[12] We have used a simple two-state model that assumes that the motion of the labeled proteins in membranes can be represented by a strongly (S) and a weakly (W) immobilized state to analyze the V_1 spectra. The S-state gives the broad-line signal and the W-state gives the narrow-line signal (each characterized by its peak-to-peak line width). At 20°C, membranes in 5 mM phosphate buffer at pH 8 exhibit a spectrum that consists of about 10% W and 90% S state. A 1% change in the states (to 9% W and 91% S, or 11% W and 89% S) will give an 11% change in the W/S value. A 5% change in the states produces a change in the W/S value of more than 50%. Therefore, even though the W component is only a minor portion of the total signal, the W/S is very sensitive to the intensity change in the W component. Small alterations in the physical state of the membrane, for example, caused by changes in temperature or pH of the sample, will cause only minor changes in the motion of the membrane proteins. However, these minor changes in protein mobility correspond to a significant change in W/S value. These studies also indicate that, if one wants to compare the W/S ratios of different samples, the sample conditions as well as EPR measurement parameters, such as temperature, must be very carefully controlled.

It should be pointed out that this simple two-state model may include a more detailed description of membrane dynamics where the S-state or the W-state consists of more than one state, as long as the substates in the S- or W-state give strongly or weakly immobilized EPR signals, respectively. Recently, Rifkind and co-workers observed an increase in EPR signal intensity upon increase in temperature and suggested that there exists a state that is EPR silent at low temperature because of dipolar interaction but becomes EPR active at higher temperatures.[24] They suggested that a model of at least three states is more appropriate. We disagree with this interpretation. We have obtained EPR spectra of Mal-6 alkylated to membrane as well as Mal-6 in phosphate-buffered saline (PBS) solution as a function of temperature. Double integration of these EPR signals shows increase in intensity in both samples (FIGURE 2). It is generally known that the sensitivity of the EPR cavity is a function of temperature and spin-concentration calibration should be obtained at each temperature. An intensity increase does not necessarily represent an increase in spin concentration. Thus, we believe that the simple two-state model appears to be appropriate for the analysis of the V_1 spectra.

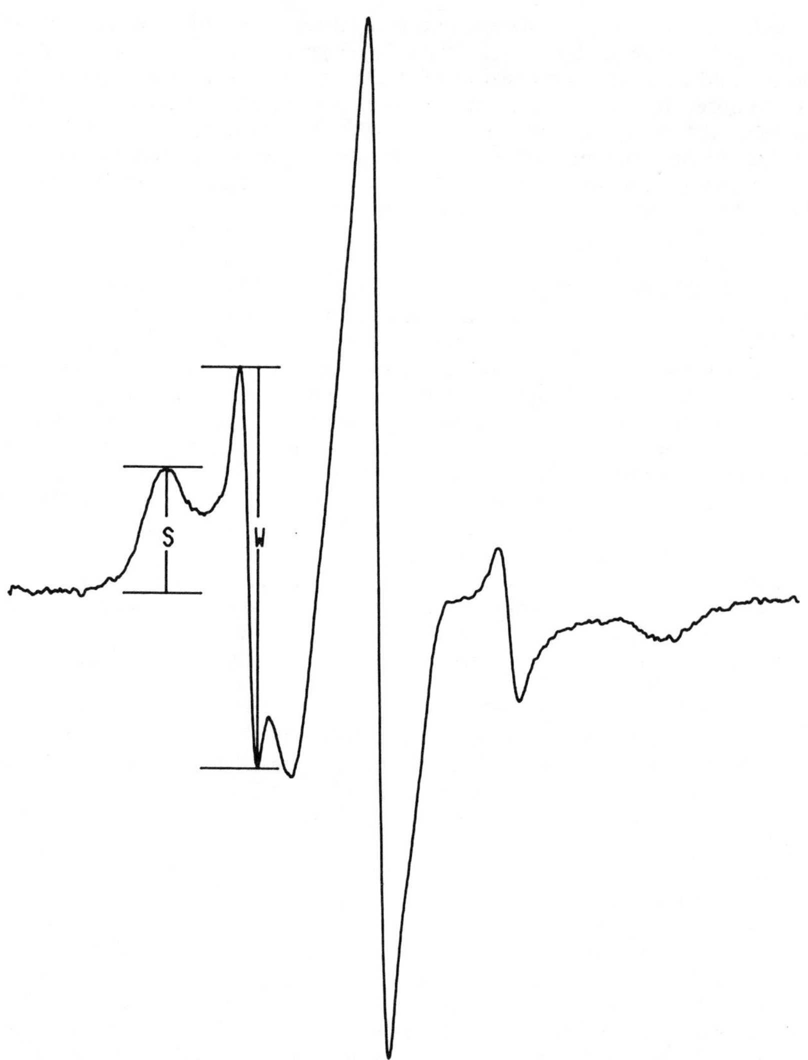

FIGURE 1. Typical conventional V_1 EPR spectrum of Mal-6–labeled erythrocyte membranes in 5 mM phosphate buffer at pH 7.4 and 20°C.

We have monitored the W/S values of membranes from patients with hereditary spherocytosis (HS) and Huntington's disease (HD). In the HS studies, the W/S ratios of normal and HS samples differed only slightly at 20°C and 40°C. However, increasing the temperature to 47°C and incubating for long time periods markedly accentuated the difference in W/S values.[13] It is suggested that the apparent differences in heat sensitivity between normal and HS membrane proteins reflect a latent structural alteration(s) of hereditary spherocytosis. In the HD studies, we did not observe differences in the W/S ratios between normal and HD erythrocyte mem-

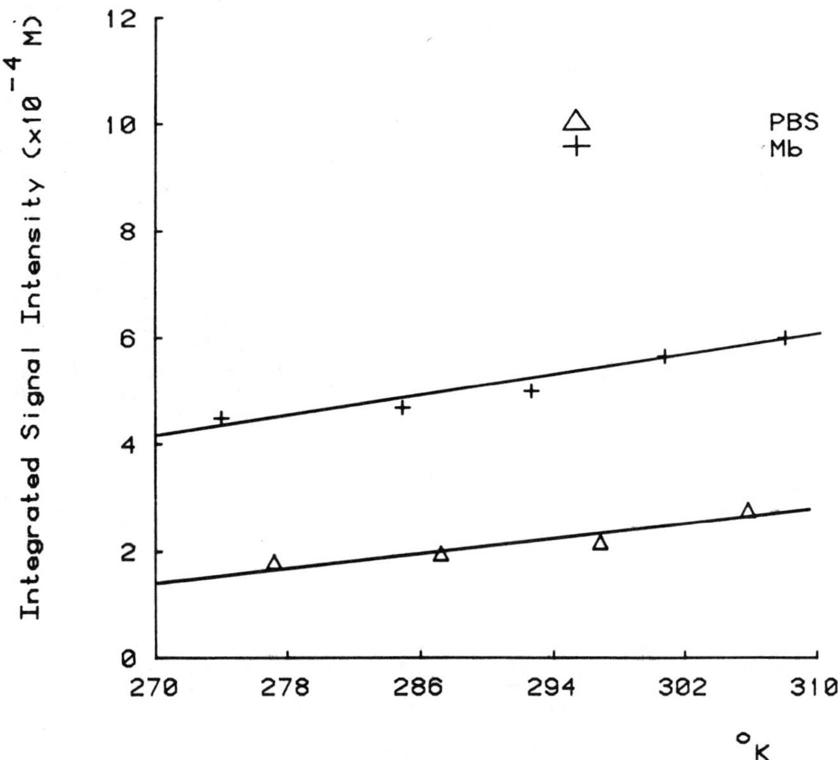

FIGURE 2. Integrated V_1 EPR signal intensities of Mal-6 alkylated to membrane proteins (Mb) (+) and in phosphate-buffered saline (PBS) (Δ) solution as a function of temperature. The temperature slopes of the two lines are statistically identical. The vertical offset between curves is arbitrary.

branes,[12] in disagreement with one of the earlier studies,[3] but in agreement with another study.[6]

In addition to the studies of erythrocyte membranes of normal and diseased individuals, we have also used the W/S values to study the hemoglobin-membrane interaction. When hemoglobin molecules interact with the cytoplasmic surface of the membrane, the motions of some of the membrane proteins are restricted, resulting in a conversion of the W-state to the S-state and thus a decrease in the W/S ratio. Due to the sensitivity of the W/S ratio, even a weak interaction can be monitored by this method. Our earlier work showed that the interaction between membrane and hemoglobin (in the *R* state) at physiological pH in 5 mM phosphate buffer was very low affinity in nature, with an equilibrium dissociation constant about 10^{-4} M.[9] Current studies compare the sickle cell and normal hemoglobin interactions with membranes. EPR measurements show that sickle cell hemoglobin causes a larger decrease in the W/S value of membrane than does normal hemoglobin.[11] Furthermore, the studies show that removal of spectrin-actin complex by low ionic strength buffer causes little change in the equilibrium dissociation constant, indicating that the remaining membrane proteins play the primary role in hemoglobin-membrane interaction. The

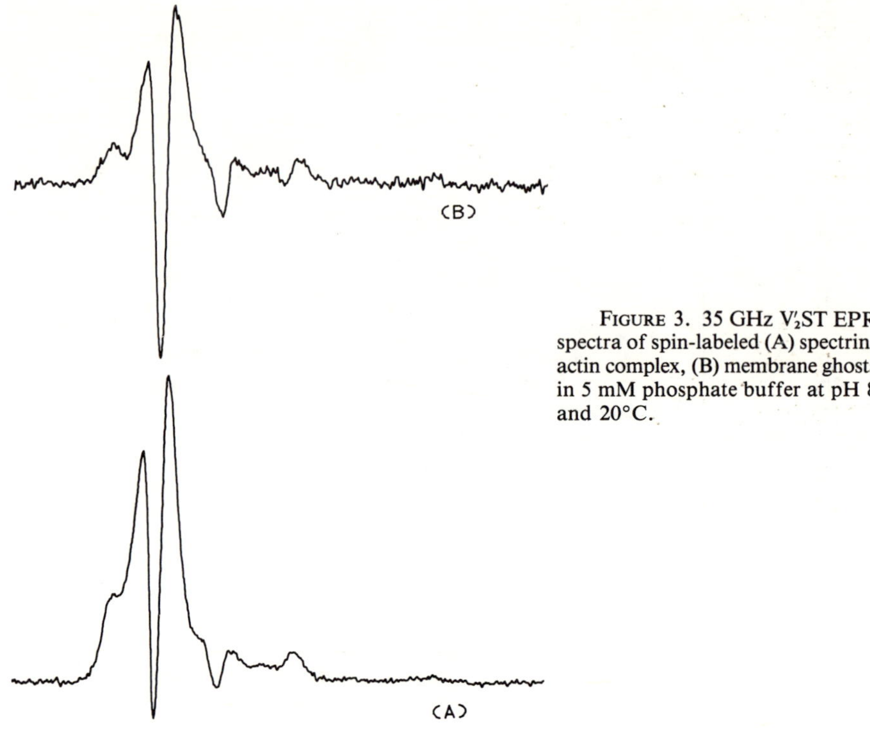

FIGURE 3. 35 GHz V_2'ST EPR spectra of spin-labeled (A) spectrin-actin complex, (B) membrane ghosts in 5 mM phosphate buffer at pH 8 and 20°C.

stronger interaction of sickle cell hemoglobin than normal hemoglobin with membranes may have some significance in the formation of irreversibly sickled cells.

EPR Spectral Properties of ST EPR (V_2') Detection and Their Applications

The strongly immobilized component in the V_1 detection, which is the major spectral component in the system, shows little or no change in spectral shape as rotational correlation times slow down beyond 10^{-7} sec. However, V_2' spectra in ST EPR show sensitivity in the range of 10^{-7} to 10^{-3} sec.[30] The V_2' spectra of labeled erythrocyte membranes with reasonable signal-to-noise ratio can be obtained by signal averaging.[8] The spectral parameters L''/L, C'/C, and H''/H introduced by Thomas *et al.*,[30] are generally used to monitor the dynamic behavior of the labeled systems.[17] V_2' spectra of Band III protein systems show that Band III's lateral mobility is affected by cytoskeletal proteins.[26] Data of membranes and of spectrin-actin complex extracted from labeled membranes by low ionic strength buffer reveal that the complex appears to exhibit complicated motions and that the motions are relatively independent of the lipid bilayer and other membrane components.[15] Another ST EPR study also shows that the dynamic behavior of actin-depleted spectrin molecules that are bound to ankyrin is qualitatively similar to that of spectrin dimers or tetramers in solution.[22] The motions observed in these two studies are believed to be wobbling and flexing motions in the cytoskeletal network. However, because of several experimental difficulties, detailed understanding of the dynamic properties of spectrin-actin is still not available. One of these difficulties is the superimposed signals observed at 9.5 GHz frequency. When the observational frequency is increased to 35 GHz (*Q*-band), the increased levels of the spin states and *g*-anisotropy give better resolved spectra.[18] We have applied the *Q*-band ST EPR techniques to obtain V_2' spectra of labeled spectrin-actin complex in solution and in ghost membrane (FIGURE 3).[10] Computer spectra subtraction is used to analyze the spectra. Preliminary analysis suggests the existence of multiple classes or rates of motions within the complex. The three principal motional components appear to have correlation times of about 10^{-9} sec, 10^{-7} to 10^{-6} sec, and slower than 10^{-3} sec, which are classified as fast, slow, and very slow motions, respectively. The fast component corresponds to the W-component in the V_1 detection. The slow and very slow components appear moderately sensitive to temperature (over the range 0°C to 30°C), but virtually independent of pH over the range of 6 to 8.

In summary, the maleimide spin-label EPR studies are beginning to provide us with some quantitative information on the motional behavior of erythrocyte membrane proteins. Since membrane component assembly is dynamic in nature, motional parameters are sensitive to membranous and cellular events and have been applied to study structural and functional properties of membranes from normal and abnormal cells. Future identification of the labeling sites on individual proteins and applications of modern EPR techniques to this system will certainly help us to further advance our understanding of the erythrocyte membrane on a molecular level.

REFERENCES

1. ABBOTT, R. E. & D. SCHACHTER. 1976. J. Biol. Chem. **251**: 7176–7183.
2. BUTTERFIELD, D. A. 1976. Acc. Chem. Res. **10**: 111–116.
3. BUTTERFIELD, D. A. & W. R. MARKESBERY. 1981. Life Sci. **28**: 1117–1131.
4. CARRAWAY, K. L. & B. C. SHIN. 1972. J. Biol. Chem. **247**: 2102–2108.
5. CHAPMAN, D., M. D. BARRATT & V. B. KAMAT. 1969. **173**: 154–157.
6. COMINGS, D. E., A. PEKKALA, J. R. SCHUH, P. C. KUO & S. I. CHAN. 1981. Am. J. Human Genet. **33**: 166–174.
7. FAIRBANKS, G., T. L. STECK & D. F. H. WALLACH. 1971. Biochemistry **10**: 2606–2617.
8. FUNG, L. W.-M. 1981. Biophys. J. **33**: 253–262.
9. FUNG, L. W.-M. 1981. Biochemistry **20**: 7162–7166.
10. FUNG, L. W.-M. & M. J. JOHNSON. 1982. J. Magn. Resonan. **51**: 233–244.
11. FUNG, L. W.-M., S. D. LITVIN & T. REID. 1982. Biochemistry **22**: 864–869.
12. FUNG, L. W.-M & M. S. OSTROWSKI. 1982. Am. J. Hum. Genet. **34**: 469–480.
13. FUNG, L. W.-M., M. S. OSTROWSKI, W. A. MEENA & S. SARNAIK. 1981. Life Sci. **29**: 2071–2079.
14. FUNG, L. W.-M. & M. J. SIMPSON. 1979. FEBS Lett. **108**: 269–272.
15. FUNG, L. W.-M., M. J. SOOHOO & W. A. MEENA. 1979. FEBS Lett. **105**: 379–383.
16. GIOTTA, G. L. & H. H. WANG. 1973. Biochim. Biophys. Acta **298**: 986–994.
17. HYDE, J. & L. R. DALTON. 1979. Saturation-transfer spectroscopy. *In* Spin Labeling: Theory and Applications. L. J. Berliner, Ed. **2**: 1–70. Academic Press, Inc. New York.
18. JOHNSON, M. E. & J. HYDE. 1981. Biochemistry **20**: 2875–2880.
19. LAMMEL, B. & G. MAIER. 1980. Biochim. Biophys. Acta **622**: 245–258.
20. LAU, P.-W., C. HUNG, K. MINAKATA, E. SCHWARTZ & T. ASAKURA. 1979. Biochim. Biophys. Acta **552**: 499–508.
21. LAURENT, M., D. DAVELOOSE, F. LETERRIER, S. FISCHER & G. SCHAPIRA. 1980. Chin. Chim. Acta **105**: 183–194.
22. LEMAIGRE-DUBREUIL, Y., Y. HENRY & R. CASSOLY. 1980. FEBS Lett. **113**: 231–234.
23. LUX, S.E. 1979. Sem. Hemotol. **16**: 21–51.
24. RIFKIND, J. M., J. G. MOHANTY, J. T. WANG, G. S. ROTH. 1982. Biophys. J. **37**: 151–152.
25. RIGAUD, J. L., C. M. GARY-BOBO & C. TAUPIN. 1974. Biochim. Biophys. Acta **373**: 211–223.

26. Sakali, J., A. Tsuji, C. H. Chang & S. I. Ohnishi. 1982. Biochemistry **21**: 2366–2372.
27. Sandberg, H. E., R. G. Bryant & L. H. Piette. 1969. Arch. Biochem. Biophys. **133**: 144–152.
28. Schneider, H. & I. C. P. Smith. 1970. Biochim. Biophys. Acta **219**: 73–80.
29. Steck, T. L. 1974. J. Cell Biol. **62**: 1–19.
30. Thomas, D. D., L. R. Dalton & J. S. Hyde. 1976. J. Chem. Phys. **65**: 3006–3024.

INDUCED ALTERATIONS IN THE PHYSICAL STATE OF SIALIC ACID AND MEMBRANE PROTEINS IN HUMAN ERYTHROCYTE GHOSTS: IMPLICATIONS FOR THE TOPOLOGY OF THE MAJOR SIALOGLYCOPROTEIN*

D. Allan Butterfield, Bennett T. Farmer II,† and Jimmy B. Feix‡

Department of Chemistry
University of Kentucky
Lexington, Kentucky 40506

Introduction

While much is known of the topology of membrane proteins in human erythrocyte membranes[1-5] schematically represented in FIGURE 1, the spatial relationships of the major sialoglycoprotein, glycophorin A, to the cytoskeleton remain unknown. Based on morphologic investigations of the distribution of colloidal iron hydroxide particles (which are solely bound to sialic acid residues, 70% of which are on glycophorin A[5]) upon incubation of human erythrocyte membranes with antispectrin antibodies, Nicholson and Painter[6] have suggested a structural linkage between the inner surface peripheral protein spectrin and the integral transmembrane protein glycophorin A. Such linkages, if generally true of membranes, may be crucially important in the physiology of cells. For example, the phenomenon of transmembrane signaling, whereby the binding of a molecule such as a hormone, neurotransmitter, or psychotropic drug to one side of the membrane results in information transfer to the opposite side of the membrane, often expressed by the action of "second messengers" on membrane, cytoplasmic, or nuclear proteins, is important in cellular regulation.[7,8] The mechanisms by which information transfer occurs through the membrane are not completely understood, but conformational alterations of proteins on the opposite side of the membrane from that at which a binding event or protein modification occurs would probably be important. In the erythrocyte membrane at least two transmembrane linkages have been demonstrated: (1) some molecules of Band 3, the major transmembrane protein, are bound to Band 2.1, which in turn is linked to the principal cytoskeletal protein spectrin[9,10] and (2) the minor glycoprotein, glycophorin C, is reported to be linked to Band 4.1,[1] a key cytoskeletal protein that is itself coupled to Band 3 and spectrin.[1] In order to investigate whether binding events or protein modification on one side of the membrane would be manifested by changes in the physical state of membrane proteins on the opposite side of the membrane and to gain further insight into the putative linkage of glycophorin A to spectrin, we have performed several series of preliminary spin-labeling experiments utilizing sialic acid–specific and membrane protein–specific spin-labeling methods.

* Supported in part by research grants from the National Institutes of Health (AG-0084 and AG-02759).

† National Institutes of Health Predoctoral Trainee through the Sanders-Brown Research Center on Aging, University of Kentucky.

‡ Present address: National Biomedical ESR Laboratory, Medical College of Wisconsin, Milwaukee, Wisconsin 53226.

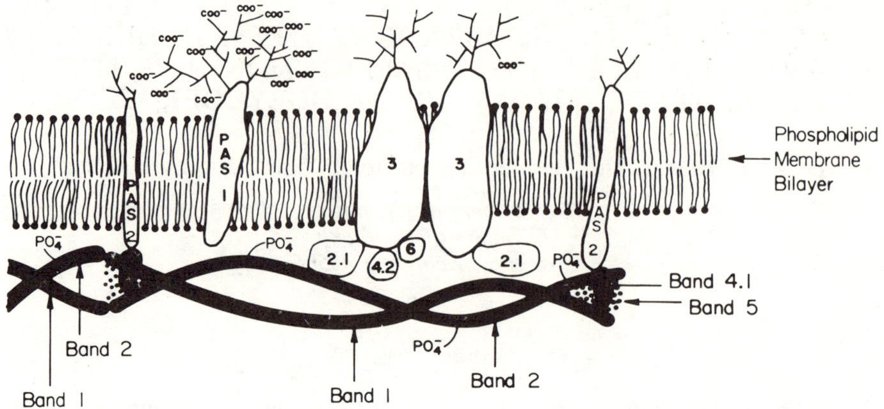

FIGURE 1. Schematic representation of protein orientations in the human erythrocyte membrane. The symbols refer to the following proteins named according to conventional nomenclature: 3, Band 3, the major transmembrane protein and anion-transporting protein; PAS-1, glycophorin A, the major sialoglycoprotein; PAS-2, minor sialoglycoprotein thought to be bound to Band 4.1; Bands 1 and 2, spectrin, the major component of the erythrocyte cytoskeleton, which also includes Band 5 (actin), Band 2.1 (the binding site of spectrin to the membrane), Band 4.1; and 4.2.[1] Band 6 is glyceraldehyde 3-phosphate dehydrogenase.[1] The configuration of Band 3 (a dimer) with respect to the number of times it penetrates the bilayer has not been definitely indicated since this assignment is still controversial.[47,48]

MATERIALS AND METHODS

The spin labels employed, 2,2,6,6,-tetramethyl-4-aminopiperidine-1-oxyl (TEMPAMINE) and 2,2,6,6-tetramethyl-4-maleimidopiperidine-1-oxyl (MAL-6), were purchased from Aldrich and Syva, respectively. Ultraviolet and melting point analyses demonstrated that no isomaleimide contamination of MAL-6 was present[11] and mass spectral, NMR, and IR analyses demonstrated that TEMPAMINE was pure.[12] *Phaseolus vulgaris* phytohemagglutinin (PHA) was obtained from Miles-Yeda, wheat germ agglutinin (WGA) from Calbiochem, and concanavalin A (Con A) from Sigma. *N*-ethylmaleimide (> 99% pure) was obtained from Aldrich and glutaraldehyde was obtained from Sigma. All other chemicals were of the highest purity obtainable.

Blood was drawn into heparinized tubes by venipuncture from healthy volunteers. Intact cells were obtained within one hour of drawing blood as described previously.[12] Depending on the experiments to be performed, erythrocyte membranes (ghosts) essentially free of hemoglobin were obtained by osmotic lysis in 5 mM sodium phosphate buffer, pH 8.0 (5P8) or 10 mM Tris-HC1, pH, 7.4 at 4°C and subsequent washings in the respective buffer. In either case, the milky white ghosts were then washed two times with 5P8 prior to the initiation of spin-labeling procedures. After overnight spin labeling of sialic residues with TEMPAMINE as described below, 10 mM Tris HC1 pH 7.4 was used to remove excess spin label from the system. No ESR signal was detectable in the supernatant of the sixth Tris wash. Estimation of membrane protein content was performed according to Lowry *et al.*[13] Measurement of membrane-bound acetylcholinesterase activity was performed as described by Ellman *et al.*[14] as modified by Steck and Kant.[15] SDS-polyacrylamide gel electro-

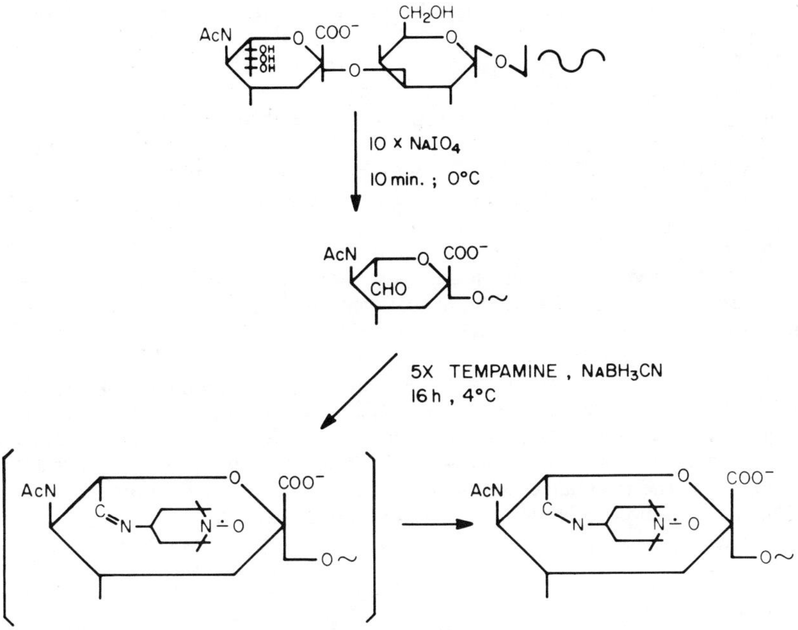

FIGURE 2. Schematic representation of the reductive amination procedure for the covalent incorporation of TEMPAMINE selectively into sialic acid residues of human erythrocyte membranes. See Feix and Butterfield[12] for more complete details.

phoresis was performed using the discontinuous buffer system of Laemmli under reducing conditions.[16]

Sialic acid in erythrocyte ghosts was spin-labeled by TEMPAMINE employing reductive amination in the presence of $NaCNBH_3$ of previously periodate-treated membranes as earlier described[12] and as outlined in FIGURE 2. No nonspecific incorporation of the spin label was observed under the conditions used for spin labeling.[12] Approximately 40% of the sialic acid residues of human erythrocytes are labeled by this method, with approximately 70% of these labeled sites on glycoproteins (most prominently, glycophorin A), and the remainder on glycolipids.[12]

Erythrocyte ghosts were spin-labeled by the protein-specific spin label, MAL-6, as described previously,[17] with care being observed to keep buffer pH, ionic strength, labeling time, and spin-label concentration constant in each experiment.[18] Lipid isolation[19] or exhaustive dialysis[20] have demonstrated that nonspecific incorporation of MAL-6 in the lipid phase of erythrocyte membranes does not occur.

In the lectin experiments, ghosts spin-labeled with TEMPAMINE or MAL-6 were diluted to 3.0 mg membrane protein per ml and the desired concentration of lectin in 5 mM sodium phosphate buffer pH 8.0 (5P8) was added at room temperature (final ghost concentration 2.0 mg protein per ml) for one hour prior to recording the electron spin resonance (ESR) spectrum. PHA was added at 1.0 mg/ml, while WGA and Con A were added at 1.5 mg/ml and 0.8 mg/ml, respectively. Additionally, Con A was studied at 1.6 mg/ml. In each case a control sample (addition of only the 5P8 buffer) was used for each sample studied. In all other experiments employing TEMPAMINE, a final spin-labeled ghost concentration of 2.0 mg/ml was used.

Glutaraldehyde was dissolved in 5P8 and added to spin-labeled ghosts to give a final concentration of 0.1%.

Spectrin was removed from erythrocyte membranes by overnight dialysis against 0.2 mM EDTA, pH 7.4 at 4°C.[21] The dialyzed samples were washed twice in the 0.2 mM EDTA, pH 7.4 buffer and twice in 5P8 and resuspended to their original volume in 5P8. Spectrin-depleted membranes were spin-labeled with TEMPAMINE as described above.

In order to study the effect of Ca^{2+} on the physical state of sialic acid, Tris-HCl washed, TEMPAMINE spin-labeled ghosts were reacted with $CaCl_2$ to give a final concentration of 10 mM. 200 μl aliquots of spin-labeled ghosts were reacted at room temperature 10 minutes with 100 μl of 10 mM Tris HCl, pH 7.4 buffer or $CaCl_2$ in 10 mM Tris HCl, pH 7.4.

Magnetic resonance spectra were obtained at room temperature on a Varian E-109 ESR spectrometer equipped with an E-238 rectangular resonant cavity and associated quartz rectangular sample cell. The waveguide and cavity were purged with extra dry nitrogen at a flow rate of five cubic feet per hour. Modulation and power broadening of the resonance lines were avoided. Under the conditions of labeling used in these studies no Heisenberg spin exchange occurred.[12]

Results

ESR Spectra of TEMPAMINE and MAL-6 Attached to Erythrocyte Membranes

In human erythrocyte membranes sialic acid is specifically spin-labeled with TEMPAMINE by reductive amination[12] as described in Methods to yield an ESR spectrum like that in Figure 3. This asymmetrically broadened three-line spectrum is characterized by an apparent rotational correlation time, τ_A, calculated by

$$\tau_A = 6.5 \times 10^{-10}\, \Delta H_{pp} \left[\sqrt{\frac{A(0)}{A(-1)}} + \sqrt{\frac{A(0)}{A(+1)}} - 2 \right],$$

where ΔH_{pp}, A(0), A(+1), and A(−1) are, respectively, the peak-to-peak linewidth of the central ($M_I = 0$) resonance line in Gauss and the peak-to-peak amplitudes of the $M_I = 0$, +1, and −1 lines.[22] The longer the τ_A, the slower the motion of spin-labeled sialic acid. Rigorous calculations of the rotational correlation time would require computer simulation of experimental spectra, inclusion of M_I-dependent broadening of the three nitrogen hyperfine lines by unresolved hydrogen hyperfine coupling,[23] and other small effects. Consequently, the absolute accuracy of the apparent correlation time calculated by the above equation is subject to the errors stated; however, differences in correlation times induced by various treatment are given precisely. Consequently, relevant biochemical information is obtainable from these systems without these further refinements.

In contrast to the three-line ESR spectrum of TEMPAMINE bound to sialic acid in erythrocyte membranes, ghost proteins spin-labeled by MAL-6 give rise to a composite spectrum indicative of at least two different classes of label-binding sites characterized by their motion: strongly and weakly immobilized (Figure 4). The maleimide label is thought to bind principally to SH groups on cysteine but it is also reported that a small proportion of the signal intensity is caused by MAL-6 incorporation onto NH_2 groups on lysine residues.[24] Spectrin, actin, Band 3, Band 2.1, and glycophorin are thought to be the chief binding sites of MAL-6 in erythrocyte ghosts

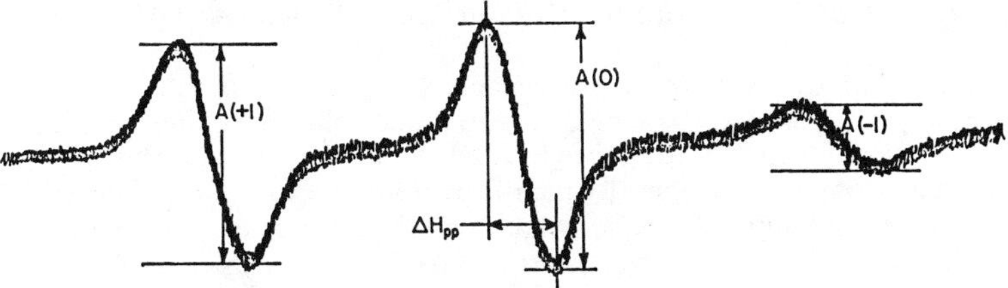

FIGURE 3. Typical ESR spectrum of TEMPAMINE spin label attached to sialic acid residues of membrane glycoproteins and glycolipids in human erythrocyte ghosts. The spin label was covalently attached to sialic acid by reductive amination as previously described.[12] The spectral measurements required in the calculation of τ_A are indicated. Instrument parameters: 40 G scan width, 14 mw microwave power and 0.32 Gauss modulation amplitude.

spin-labeled as described above.[25] Other scientists, using different spin-labeling methods report that the spectrin-actin complex accounts for 80% of the MAL-6 signal intensity, Band 3 for 10%, while the binding sites for the remaining 10% are unknown.[26] The MAL-6 spectrum of erythrocyte ghosts has been discussed extensively elsewhere[19,27] as have the several experimental variables that need to be controlled to give reproducible ESR spectra employing this label.[18] The analysis of the spectrum of MAL-6 attached to membrane proteins in erythrocyte ghosts is performed via the W/S ratio (FIGURE 4), a sensitive and convenient monitor of the physical state of membrane proteins.[18,19,27-30] The W/S ratio is reported to be altered by various procedures and agents that change the conformation of membrane proteins.[27-30]

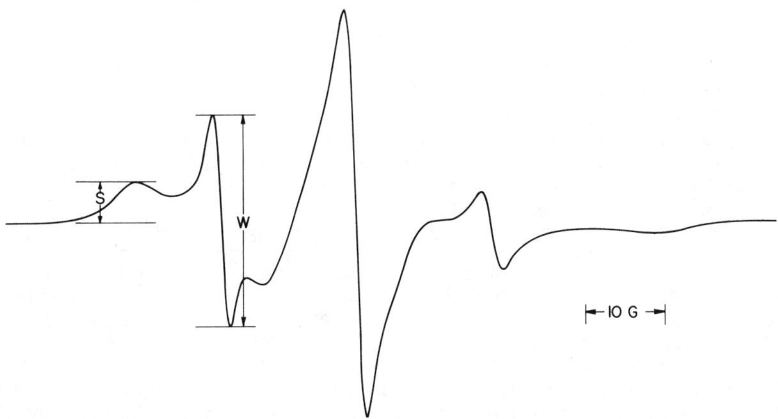

FIGURE 4. A typical ESR spectrum of MAL-6 attached to membrane proteins in normal erythrocyte membranes. The amplitudes of the $M_I = +1$ lines of the spin label covalently bound to strongly and weakly immobilized binding sites are indicated by S and W, respectively. Fresh erythrocyte ghosts were labeled overnight at 4°C by MAL-6 in 5P8 as previously described.[17] After five washes with 5P8 to remove excess spin label, the spectra of the spin-labeled ghosts were recorded at room temperature with a sweep width of 100 G. The modulation amplitude and microwave power incident on the resonant cavity were 0.32 G and 14 mw, respectively.

Interaction of Lectins with the Cell Surface of Spin-labeled Membranes

Lectins, plant or invertebrate proteins or glycoproteins, bind selectively to specific carbohydrate residues. PHA binds to red blood cell membranes via strong interaction with glycophorin A.[31] Mannose located in the core of the carbohydrate chain and two galactose residues, one of which is penultimate to a sialic acid residue, are the primary determinants. *P. vulgaris* PHA binding is not affected by removal of sialic acid.[32] The erythrocyte receptor for WGA is also glycophorin A.[33] However, several studies suggest that in addition to *N*-acetylglucosamine, sialic acid is also involved in the binding of this lectin.[33-35] Con A is thought to bind to Band 3,[36] which contains very little sialic acid,[5] via glucopyranosides and mannopyranosides.

The results of the effect of the three lectins PHA, WGA, and Con A on τ_A of TEMPAMINE and the W/S ratio of MAL-6 are given in TABLE 1. PHA and WGA, which both bind to glycophorin A but at different sites (discussed above), gave a 10% and 33% mean reduction in the apparent rotational correlation time of the spin label attached to sialic acid (TABLE 1). In contrast, Con A, which binds to Band 3, did not affect τ_A of TEMPAMINE in human erythrocyte ghosts. The W/S ratio of the protein-specific spin-label MAL-6 is highly significantly reduced by PHA, increased by WGA, and unaffected by Con A (TABLE 1), suggesting that the former two lectins, though both binding at the cell surface to glycophorin, may have pronounced but markedly different effects on the conformation of proteins within the membrane.

Effects of Modification of Membrane Proteins on τ_A of TEMPAMINE Bound to Cell Surface Sialic Acid

The results of lectin binding monitored by the protein-specific spin-label MAL-6 (TABLE I) suggested that binding events occurring at the cell surface may be manifested in changes in the physical state of proteins within the membrane. We wished

TABLE 1

EFFECTS OF THE LECTINS PHA, WGA, AND CON A ON τ_A OF THE SIALIC ACID-SPECIFIC SPIN LABEL TEMPAMINE AND THE W/S RATIO OF THE PROTEIN-SPECIFIC SPIN LABEL MAL-6 IN HUMAN ERYTHROCYTE GHOSTS*

Lectin	Concentration (mg/ml)	Protein To Which Lectin Is Bound	*N*	$\tau_A/\tau_{control}$	$(W/S)_{lectin}/(W/S)_{control}$
PHA	1.0	Glycophorin	8	0.90 ± 0.03†	0.67 ± 0.03†
WGA	1.5	Glycophorin	4	0.67 ± 0.13‡	1.17 ± 0.05§
Con A	0.8	Band 3	3	0.99 ± 0.02	1.00 ± 0.04
	1.6	Band 3	3	1.00 ± 0.11	–

* A separate control sample was obtained for each individual lectin sample studied on each day. The level of significance (p value) of each mean ratio was calculated by a two-tailed Student's *t*-test with a null hypothesis that the respective ratio is unity (the value expected if the particular lectin under study were to cause no alteration in the apparent correlation time or physical state of membrane proteins compared to control) and an alternative hypothesis that this ratio is different than unity. Mean ± Std. Dev. are presented.

† $p < 0.0001$.

‡ $p < 0.02$.

§ $p < 0.05$.

TABLE 2

EFFECTS OF MEMBRANE PROTEIN MODIFICATION IN HUMAN ERYTHROCYTES GHOSTS ON THE APPARENT ROTATIONAL CORRELATION TIME OF SPIN-LABELED SIALIC ACID RELATIVE TO UNTREATED CONTROLS*

Agent Added	Effect Of Agent On Membrane Proteins	$(\tau_A)_{modified} / (\tau_A)_{control}$	*N*	p†
0.4 mM NEM	SH groups chemically modified	1.01 ± 0.04	5	N.S.
0.1% Glutaraldehyde	Reported to cross-link spectrin without cross-linking band 3 or glycophorin	1.21 ± 0.10	4	<0.05

* Mean ± Std. Dev. are presented.
† Value of p calculated by a two-tailed Student's *t*-test.

to determine whether modification of membrane proteins within the bilayer or attached to the cytoplasmic surface would result in alterations in the physical state of sialic acid on the external side of the membrane.

TABLE 2 demonstrates that with simple chemical modification of membrane proteins by *N*-ethylmaleimide, no alteration in τ_A of spin-labeled sialic acid occurred. However, when spin-labeled ghosts were treated with 0.1% glutaraldehyde, a concentration that is reported by Steck[37] to cross-link spectrin on the cytoplasmic side of the membrane without cross-linking Band 3 or glycophorin A, a significant ($p < 0.05$), 21% reduction in rotational motion of spin-labeled sialic acid was observed. This result may mean that spectrin and glycophorin A interact closely in the membrane and does suggest that modification of a protein on the cytoplasmic side of membrane can be reflected in alteration of the physical state of the external side. Acetylcholinesterase activity in erythrocytes treated with 0.1% glutaraldehyde was the same as normal controls (data not shown) consistent with the suggestion that spectrin is the principal protein affected by the low concentration of glutaraldehyde used.[37]

Additional preliminary evidence of the importance of spectrin in the conformation of glycophorin has been obtained. Previous spin-labeling studies utilizing MAL-6 had shown that removal of spectrin from the membrane resulted in conformational changes of large magnitude in the remaining membrane proteins.[18] The present work has suggested that the rotational motion of spin-labeled sialic acid of ghosts depleted of spectrin was 87% slower than that of control ghosts (TABLE 3, $p < 0.02$). It is not clear from these experiments alone whether this finding is a consequence of the absence of spectrin-glycophorin A interactions, a secondary result of the effect of spectrin removal on the physical state of other membrane proteins which then causes the decreased motion of sialic acid, or some combination of these two. However, the initial results are consistent with the idea that spectrin may have considerable influence on the physical state of sialic acid and that alterations in the conformation of proteins within the membrane caused by the removal of spectrin from the cytoplasmic surface may be expressed at the external surface of the membrane.

Further studies designed to determine if membrane modification would affect the physical state of sialic acid were undertaken. Calcium is known to have a notable effect on the physical state of membrane proteins in erythrocyte ghosts[18,38] and to bind to spectrin.[39] However, Ca^{2+} is also reported to interact with negatively charged lipids in vesicles to cause a phase separation.[40] In erythrocyte membranes the most

TABLE 3

COMPARISON OF THE APPARENT ROTATIONAL CORRELATION TIME OF SPIN-LABELED SIALIC ACID IN SPECTRIN-DEPLETED HUMAN ERYTHROCYTE GHOSTS TO THAT OF INTACT GHOST MEMBRANES

Experiment		$\frac{(\tau_A)_{spectrin\text{-}depleted}}{(\tau_A)_{control}}$
1		1.80
2		1.72
3		2.09
	Mean ± S.E.M.	1.87 ± 0.11
	p*	<0.02

* Value of p calculated by a two-tailed Student's *t*-test with a null hypothesis that the mean ratio was 1.00, the value expected if spectrin depletion had no effect on the motion of spin-labeled sialic acid, and an alternative hypothesis that this mean ratio was not unity.

abundant negatively charged lipid is phosphatidylserine, which is thought to be found predominantly on the inner (cytoplasmic) aspect of the lipid bilayer.[1-5] In our initial studies, Ca^{2+} at 10 mM in 10 mM Tris HC1, pH 7.4 was incubated with TEMPAMINE spin-labeled ghosts. A 20% decrease in τ_A (increased motion) was observed (TABLE 4, $p < 0.005$).

DISCUSSION

These spin-label studies have suggested that modification of membrane proteins on one side of the membrane may result in an alteration in the physical state of proteins on the opposite side of the bilayer. In particular, lectin binding to glycophorin A at the external cell surface caused significant changes in the conformation of proteins within the membrane. Similarly, alterations in the physical state of proteins within the membrane, particularly spectrin at the cytoplasmic side of the membrane, were capable of altering the physical state of sialic acid on the external side of the membrane. The interaction of three cell-surface–specific lectins, two of which bind to different sites on the same glycoprotein, can be discriminated by both sialic acid and protein-specific spin labels. PHA caused a mean increase in motion of spin-labeled sialic acid of 10% while decreasing the W/S ratio, a measure of the conformation of proteins in the membrane, by greater than 30%. In contrast, WGA increases the motion of spin-labeled sialic acid by more than 30% while increasing the W/S ratio of spin-labeled proteins in erythrocyte membranes by approximately 17%.

The simplest interpretation of the results with PHA or WGA on the motion of sialic acid is that binding of these lectins may induce a conformational change in the glycoprotein receptor, which exposes the spin label to a less restricted environment, allowing the rotational mobility of the TEMPAMINE population to increase and to a greater extent with WGA than PHA. The MAL-6 results show that PHA and WGA binding to glycophorin A at the external surface of human erythrocyte membranes have pronounced but different effects on the physical state of proteins within the membrane.

Our experiments with Con A, which binds to Band 3, suggest that treatment of spin-labeled ghosts with this lectin induces no change in the mobility of the sialic acid–bound probe or in the conformation of proteins in the membrane (TABLE 1). Con A is also reported not to alter human erythrocyte membrane fluidity.[41] Since

TABLE 4

COMPARISON OF THE EFFECT OF 10 mM Ca^{2+} ON THE APPARENT ROTATIONAL CORRELATION TIME OF SPIN-LABELED SIALIC ACID IN ERYTHROCYTE GHOSTS RELATIVE TO CONTROL

Experiment		$\frac{(\tau_A)_{Ca^{2+}}}{(\tau_A)_{buffer}}$
1		0.70
2		0.66
3		0.86
4		0.84
5		0.89
6		0.86
	Mean ± S.E.M.	0.80 ± 0.04
	p*	<0.005

* Value of p calculated by the two-tailed Student's *t*-test with null hypothesis that the mean ratio is 1.00, the value expected if 10 mM Ca^{2+} had no effect on the motion of spin-labeled sialic acid, and the alternative hypothesis that Ca^{2+} did affect motion of sialic acid.

Con A has a binding specificity for glucopyranosides and mannopyranosides, it is unlikely that our spin labels have prevented its attachment to the membrane.

No alterations in τ_A were observed upon simple chemical modification of membrane proteins with *N*-ethylmaleimide. However, cross-linking of spectrin with 0.1% glutaraldehyde is sufficient to significantly reduce the motion of sialic acid (TABLE 2). Glutaraldehyde at this low concentration is also sufficient to decrease erythrocyte lipid fluidity.[42] But this result is suggested to be a secondary consequence of protein alterations.[43] Consequently, while lipid alterations induced by glutaraldehyde could be partially responsible for the decreased motion of spin-labeled sialic acid, spectrin alterations may possibly be more important. Removal of spectrin from the membrane decreased the molecular motion of sialic acid even more than partial cross-linking of this protein (TABLE 3).

Glycophorin contains over 70% of the red blood cell sialic acid[5] and incorporates the greatest proportion of TEMPAMINE.[12] While the concept of "boundary" lipids is still controversial,[1-5] one ^{31}P-NMR study suggests that glycophorin A immobilizes phospholipids at the glycophorin-lipid interface.[44] The results of the present preliminary studies suggest the possibility that glycophorin A and spectrin are linked in the membrane in a dynamic tension. Perturbations to spectrin, which partially cross-link this protein, cause decreased motion of sialic acid; however, as noted, this motion is decreased even further if spectrin is removed. Under these conditions, phospholipid-glycophorin A interactions may possibly increase resulting in decreased motion. Concomitantly or alternatively, removal of spectrin would diminish Band 3–spectrin interactions through Band 2.1 and could conceivably cause the reported transient interactions of glycophorin A and Band 3[45] to become stronger and longer in duration, thereby possibly resulting in decreased motion of sialic acid. Additional studies utilizing purified components will be required to begin to address these possibilities. The findings of our studies suggesting that spectrin has great influence on the physical state of sialic acid is consistent with and supportive of the morphological studies of Nicholson and Painter,[6] who suggest a direct link between spectrin and glycophorin. Repetition of the experiments described in the present study utilizing erythrocytes from persons with the rare blood type En(a –), who lack glycophorin A but have sialic acid on Band 3,[49] would provide a means to further test this suggestion.

The molecular basis for the suggested increased motion of spin-labeled sialic acid caused by Ca^{2+} (TABLE 4) is unknown. Calcium is reported to bind to spectrin.[39] In addition to the perturbation of membrane protein conformation by Ca^{2+} reported by several laboratories,[18,38,46] the observed conformation of Ca^{2+} could potentially be partially due to a relaxation of glycophorin-lipid interactions induced by this divalent cation through its binding to negatively charged phospholipids, usually located on the cytoplasmic surface of the cell. Further investigations utilizing purified glycophorin A are needed to assess the degree to which these two possibilities are important in our results.

The present initial studies provide evidence for the concept that protein conformational changes on one side of the membrane can be manifested by protein alterations on the opposite side of the bilayer (TABLES 1-4). These investigations also provide experimental support for the role of transmembrane protein conformational alterations in mechanisms of information transfer. A binding event on one side of the bilayer may cause a protein conformational change on the opposite side of the membrane, which in turn may initiate or regulate subsequent metabolic, binding, or chemical events thereby completing the information transfer. Experimental testing of these ideas are possible with the spin-labeling systems described in this study.

REFERENCES

1. MORRISON, M., T. J. MUELLER & H. H. EDWARDS. 1981. Protein architecture of the erythrocyte membrane. *In* The Function of Red Blood Cells: Erythrocyte Pathobiology. D. F. H. Wallach, Ed.: 17–34. Alan R. Liss. New York.
2. LUX, S. E. 1979. Sem. Hematol. **16**: 21–81.
3. RICE-EVANS, C. & D. CHAPMAN. 1981. Scand. J. Clin. Lab. Invest. **41**(Suppl. 156): 99–110.
4. PALEK, J. 1981. Texas Rep. Biol. Med. **40**: 397–416.
5. MARCHESI, V. T. 1979. Sem. Hematol. **16**: 3–20.
6. NICHOLSON, G. L. & R. G. PAINTER. 1973. J. Cell Biol. **59**: 395–406.
7. NATHANSON, J. A. & P. GREENGARD. 1977. Sci. Am. **237**: 108–119.
8. BRAMHALL, J., B. ISHIDA & B. WISNIESKI. 1978. J. Supramol. Struct. **9**: 399–406.
9. BENNETT, V. & P. J. STENBUCK. 1979. Nature **280**: 468–473.
10. BENNETT, V. & P. J. STENBUCK. 1980. J. Biol. Chem. **255**: 6424–6432.
11. BARRATT, M. D., A. P. DAVIES & N. T. A. EVANS. 1971. Eur. J. Biochem. **24**: 280–283.
12. FEIX, J. B. & D. A. BUTTERFIELD. 1980. FEBS Lett. **115**: 185–188.
13. LOWRY, O. H., N. J. ROSEBROUGH, A. L. FARR & R. J. RANDALL. 1951. J. Biol. Chem. **193**: 265–275.
14. ELLMAN, G. L., K. D. COURTNEY, A. VALENTINO & R. M. FEATHERSTONE. 1961. Biochem. Pharmacol. **7**: 88–95.
15. STECK, T. L. & J. A. KANT. 1974. Methods Enzymol. **31**: 172–180.
16. LAEMMLI, U. K. 1970. Nature **227**: 680–685.
17. BUTTERFIELD, D. A., P. F. DOORLEY & W. R. MARKESBERY. 1980. Life Sci. **27**: 609–615.
18. BUTTERFIELD, D. A. & W. R. MARKESBERY. 1981. Biochem. Int. **3**: 517–525.
19. BUTTERFIELD, D. A. & W. R. MARKESBERY. 1981. Life Sci. **28**: 1117–1131.
20. BUTLER, K. W., N. H. TATTRIE & I. C. P. SMITH. 1974. Biochim. Biophys. Acta **363**: 357–360.
21. STECK, T. L. & J. YU. 1973. J. Supramol. Struct. **1**: 220–232.
22. NORDIO, P. L. 1976. General magnetic resonance theory. *In* Spin Labeling. Theory and Applications. L. J. Berliner, Ed.: 29–35. Academic Press. New York.
23. POGGI, G. & C. S. JOHNSON. 1970. J. Magn. Res. **3**: 436–445.
24. CHAPMAN, D., M. D. BARRATT & V. KAMAT. 1969. Biochim. Biophys. Acta **173**: 154–157.
25. BUTTERFIELD, D. A., A. D. ROSES, S. H. APPEL & D. B. CHESTNUT. 1976. Arch. Biochem. Biophys. **177**: 226–234.

26. FUNG, L. W. M. & M. J. SIMPSON. 1979. FEBS Lett. **108**: 269–273.
27. BUTTERFIELD, D. A. 1982. Spin labeling in disease. *In* Biological Magnetic Resonance. L. J. Berliner & J. Reuben, Eds. **4**: 1–78. Plenum Publishing Corp. New York.
28. HOLMES, D. E. & L. H. PIETTE. 1970. J. Pharm. Exp. Ther. **173**: 78–84.
29. KIRKPATRICK, F. H. & H. E. SANDBERG. 1973. Biochim. Biophys. Acta **298**: 209–218.
30. BUTTERFIELD, D. A., A. AKAYDIN & M. L. KOMMOR. 1981. Biochem. Biophys. Res. Commun. **102**: 190–196.
31. KORNFELD, S. & R. KORNFELD. 1969. Proc. Natl. Acad. Sci. USA **63**: 1439–1446.
32. KORNFELD, R. & S. KORNFELD. 1970. J. Biol. Chem. **249**: 2536–2545.
33. ADAIR, W. L. & S. KORNFELD. 1974. J. Biol. Chem. **249**: 4696–4704.
34. BHAVANANDAN, V. P. & A. W. KATLIC. 1979. J. Biol. Chem. **254**: 4000–4008.
35. LOVERN, R. E. & R. A. ANDERSON. 1980. J. Cell Biol. **85**: 534–548.
36. FINDLAY, J. B. C. 1974. J. Biol. Chem. **254**: 4398–4403.
37. STECK, T. L. 1972. J. Mol. Biol. **66**: 295–305.
38. ADAMS, D., M. E. MARKES, W. J. LEIVO & K. L. CARRAWAY. 1976. Biochim. Biophys. Acta **426**: 38–45.
39. BEAVEN, G. H. & W. B. GRATZER. 1980. Biochim. Biophys. Acta **600**: 140–149.
40. OHNISHI, S. I. & S. TOKOTUMI. 1981. ESR studies of calcium- and proton-induced phase separations in phosphatidylserine-phosphatidylcholine mixed membranes. *In* Biological Magnetic Resonance. L. J. Berliner & J. Reuben, Eds. **3**: 121–153. Plenum Press. New York.
41. LYLES, D. S. & F. R. LANDSBERGER. 1976. Proc. Natl. Acad. Sci. USA **73**: 3497–3501.
42. GIAVEDONI, E. B., R. P. MASON & A. P. DALMASSO. 1978. J. Immunol. **120**: 2003–2007.
43. KOMOROSKA, M., M. KOTER, G. BARTOSZ & J. GOMULKIEWICZ. 1982. Biochim. Biophys. Acta **686**: 94–98.
44. YEAGLE, P. L. & A. Y. ROMANS. 1981. Biophys. J. **33**: 243–252.
45. NIGG, E. A., C. BRON, M. GIRARDET & R. J. CHERRY. 1980. Biochemistry **19**: 1887–1893.
46. KING, L. E. & M. MORRISON. 1977. Biochim. Biophys. Acta **471**: 162–168.
47. STECK, T. L. 1978. J. Supramol. Struct. **8**: 311–324.
48. JENKINS, R. E. & M. J. A. TANNER. 1977. Biochem. J. **161**: 139–147.
49. GAHMBERG, C. G., G. TAUREN, I. VIRTANEN & J. WARTIVAARA. 1978. J. Supramol. Struct. **8**: 337–347.

BIOENERGETIC STUDIES OF CELLS WITH SPIN PROBES*

Rolf J. Mehlhorn and Lester Packer

Membrane Bioenergetics Group
Lawrence Berkeley Laboratory
and
Department of Physiology-Anatomy
University of California
Berkeley, California 94720

Introduction

Nitroxide spin labels have been used successfully for a diversity of biological studies, including studies of membrane structure,[1] protein structure,[2] and cytoplasmic water viscosity.[3] Recently another application of spin labels has evolved—the measurement of bioenergetic parameters.[4] These bioenergetic parameters are: cell volumes, pH and electrical gradients across membranes, electrical surface and boundary potentials, and one-electron reduction/oxidation potentials within specific cellular domains. The success of these measurements is due to the ease of resolving ESR spectra of nitroxide probes quantitatively to obtain membrane-bound and aqueous populations together with the availability of a host of different nitroxide structures, providing the investigator with great flexibility of probing specific cellular domains. Two basic approaches for bioenergetic measurements have been elaborated: in the first, membrane binding of certain nitroxides is exploited to relate partitioning of these probes between aqueous and membrane domains to energized states; the second consists of quantitating probe concentrations by direct measurements of spectral lines in all of the relevant cellular compartments. The former approach requires calibration curves derived from model system studies and thus closely resembles analogous experiments conducted with fluorescent probes, radioactive tracers, etc. The latter approach avoids difficulties associated with model system calibrations by virtue of performing direct measurements of probe concentrations using membrane-impermeable paramagnetic broadening agents to selectively eradicate signals from the extracellular water. Here we present the technique that uses paramagnetic broadening agents for bioenergetic measurements.

Cell Volumes

Accurate volume determinations are required for quantitative bioenergetic studies. Furthermore, volume changes occurring under energized conditions can often be used to infer ion fluxes and mechanisms of bulk ion transport. The nitroxide probe method is particularly suited for volume measurements because rapid measurements can be conducted requiring only minimal amounts of cells. Typically, accurate measurements can be obtained for sample volumes of 40 microliters containing membranes at a concentration of 10 mg/ml. The method consists of quantitating the probe signal in the cell suspension by measuring aqueous line heights of a membrane-

* Research supported by National Institutes of Health (GM-24273) and the U.S. Department of Energy, Division of Biological Energy Research, Office of Basic Energy Sciences.

0077-8923/83/0414-0180$01.75/0

permeable, yet highly water-soluble, probe in the absence and presence of a paramagnetic, impermeable quenching agent. An example of such a determination is presented in FIGURE 1.

The major shortcoming of such volume measurements is that many cells reduce nitroxides to their non-paramagnetic hydroxylamine derivatives. Such reduction can be caused by electron transport[5] or by cellular reducing substances, such as glutathione and ascorbate. To some extent this complication can be avoided by using low concentrations of ferricyanide as the quenching agent, since it maintains nitroxides in their oxidized state without significantly broadening the signal if its concentration is less than 1 millimolar. If the intracellular environment has a reducing potential while the exterior is much more oxidizing because of the ferricyanide, the effect of nitroxide reduction inside the cells can be assessed quantitatively by varying the ferricyanide concentration and by extrapolating the intracellular nitroxide signal of a very rapidly permeable probe to infinite ferricyanide concentrations. Unfortunately, some irreversible loss of nitroxide signals occurs even in the presence of ferricyanide, indeed certain destructive reactions may be promoted in its presence.[6] Such difficulties can be circumvented by conducting rapid-mix experiments with very rapidly per-

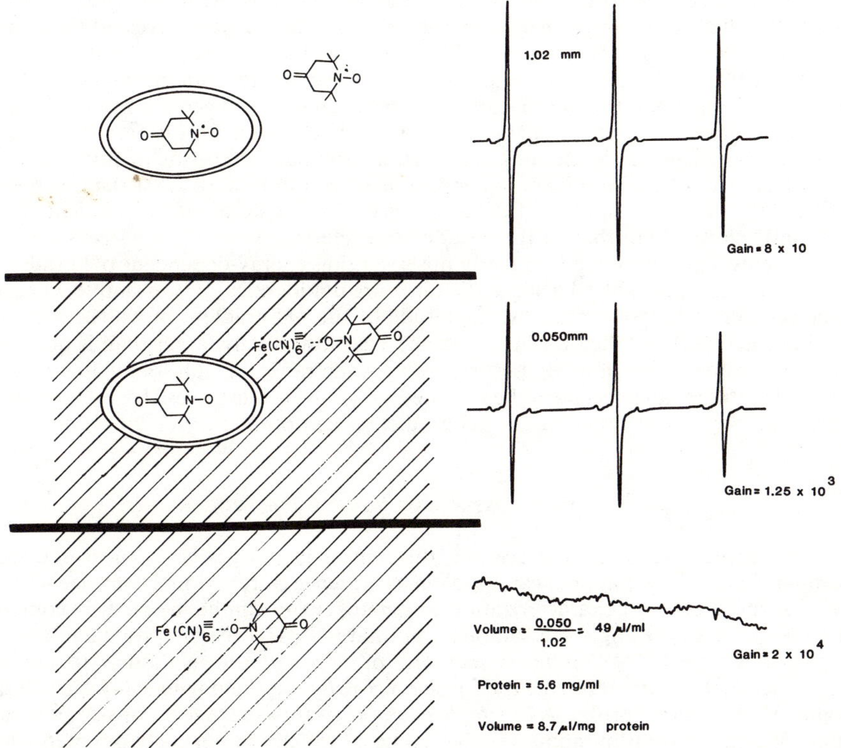

FIGURE 1. Schematic representation of a volume measurement and an example of applying the method to envelope vesicles of the *Halobacterium halobium* strain S9, 4 M NaCl, using 200 mM Na_3FeCN_6 as the quenching agent. Line heights shown for the low-field aqueous lines were divided by the respective instrument gains.

meable probes so that intracellular signals can be measured before significant signal loss has occurred. Alternatively, the rate of spin reduction can be assayed to extrapolate the intracellular signal to zero time.

To permit selection of optimally permeable probes, rates of diffusion of nitroxides across membranes must be estimated. Even with rapid-mixing techniques many probes equilibrate across all membranes too rapidly for measurement. However, such experiments do show that small nitroxides (e.g., TEMPONE) equilibrate across human erythrocyte membranes within 100 milliseconds.

pH Gradient Measurements

The technique for quantitating pH gradients with spin probes has been described.[7,8] Briefly, the method consists of performing an intracellular spin signal measurement, as described in the preceding section, but now using spin-labeled amines and weak acids. Thus, the concentration ratio of amines and acids on two sides of a membrane can be calculated and related to pH gradients. The method relies upon the assumption that the uncharged species are permeable and the charged species are not. Experiments with permanently charged spin labels of similar ionic radius have corroborated the validity of these assumptions. Moreover, experiments conducted with cells suspended in solution of varying pH have revealed that rapid permeation of the probes occurs at pH 7 (half-time of about 1 sec for uptake into red cells), but that as the pH of the solution differs increasingly from the p*K* value of the acid or amine probes, their permeability decreases and often becomes rate-limiting in measurements of kinetics of energy-dependent pH changes. This is demonstrated in terms of the different kinetics of probe accumulation and extrusion in envelope vesicles of the *Halobacterium halobium* strain S9 shown in Figure 2. At the low pH of this experiment, the response of the amine probe to the decay of the gradient is significantly slower than that of the weak acid probe.

Interestingly, we have frequently observed that energy-dependent pH gradients measured with spin-labeled acids and amines are not identical. For example, in envelope vesicles of halobacteria, the magnitude of the pH gradient measured with the acid probe (which is taken up by the vesicles) is substantially larger than is obtained with the amine. This anomaly persists over a wide range of bulk-pH values.[8] Most likely, the discrepancy between the two measurements is due to vesicle heterogeneity, i.e., some vesicles generate larger pH gradients than others.

Dual Spin-Probe Assays of Vesicle Heterogeneity

As mentioned in the last section, differences in pH gradients measured with spin-labeled acids and amines can be taken as evidence of functional heterogeneity of cells. A more detailed characterization of heterogeneity can be inferred by combining two probes in a single cell sample: a normal ^{14}N pH-sensitive probe and an isotopically enriched ^{15}N volume probe. The principle of the method is to use the pH-dependent accumulation of the former nitroxide to paramagnetically quench the signal of the volume probe. As a simple example of this approach, consider a mixed population of proton-pumping vesicles, some of which are normally oriented while the others are inverted. With an appropriate choice of probe concentrations, a decrease of the volume-probe signal will occur for vesicles where the pH-driven accumulation of the amine probe occurs, while no signal change of the ^{15}N probe will accompany extrusion from inversely oriented vesicles. Vesicles labeled with a combi-

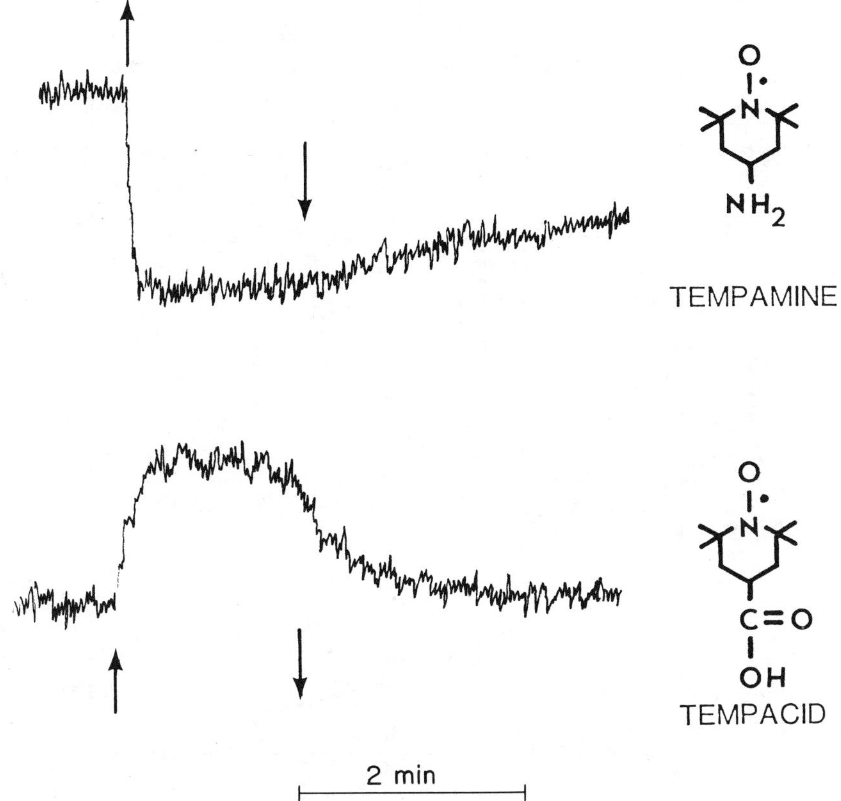

FIGURE 2. Comparison of the kinetics of ΔpH measurements with amine and weak acid probes. S9 vesicles, 3.8 M NaCl, 0.1 M Na_3FeCN_6, pH 5.3.

nation of the acid and volume probes will manifest the opposite phenomenon, i.e., only the volume probes in the inversely oriented vesicles will be quenched. Thus if one observes pH-gradient–dependent quenching of a volume probe by both acid and amine spin-labels, a mixed population of vesicles having oppositely oriented proton pumps can be inferred.

An extension of the dual-probe method is to estimate volumes of subcellular compartments whose pH differs significantly from the cytoplasm value. Such an experiment would consist of a volume determination of a cell in the presence and absence of spin-labeled amines and acids. Assuming it is known (e.g., from cytochemical staining) that such compartments exist, then the reduction of volume signals in the presence of pH-sensitive nitroxides can be used for a direct determination of the volumes of these compartments without resorting to time consuming and ambiguous cell fractionation experiments. Quenching of the signal of a volume probe by increasing concentrations of a spin-labeled amine is shown in FIGURE 3.

The permeability of available nitroxides varies considerably as exemplified by FIGURE 4. This provides the opportunity of loading cells with nitroxides by extended incubation rather than by lysis and resealing of membranes, followed by the usual

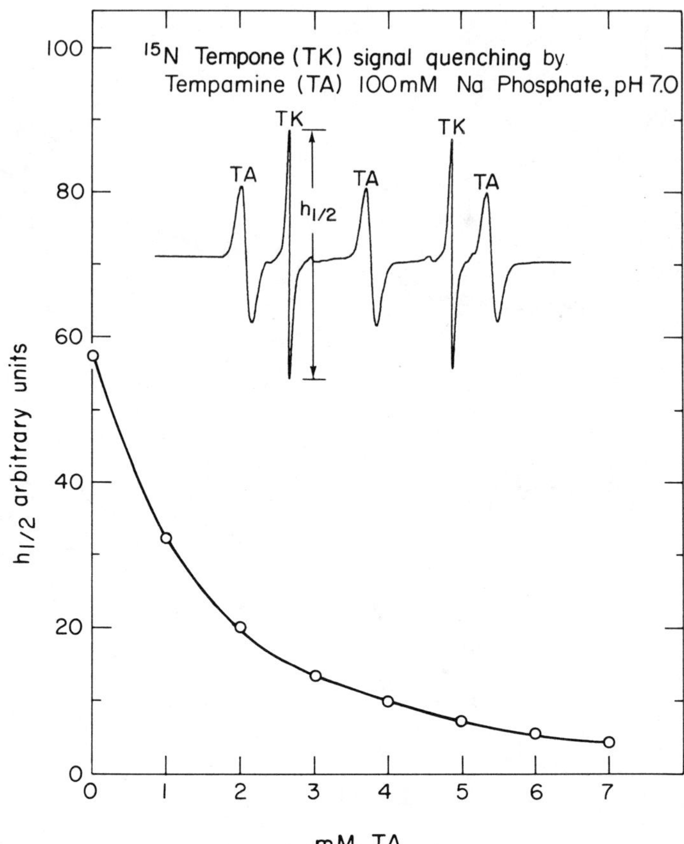

FIGURE 3. Effect of a spin-labeled amine on the low-field line height of 1 mM TEMPONE.

column chromatography to remove extracellular labels. Furthermore, such slowly permeable probes could perhaps be used for studies of cell heterogeneity. If the permeability of a probe across the plasma membrane is comparable to its permeability across subcellular membranes, then subcellular aqueous environments can be selectively observed in time-dependent experiments. Assuming that probe reduction can be avoided or corrected for, such experiments, in principle, could provide evidence for non-uniform viscosity of intracellular water, for example.

TRANSMEMBRANE ELECTRICAL POTENTIALS

Despite enormous interest in electrical potentials ($\Delta\Psi$), no truly satisfactory method has existed for accurately measuring such potentials. Microelectrodes inserted into cells necessarily break the membrane whose potential is being sought, while permeable ion accumulation suffers from uncertainties associated with membrane binding of the ions. The recent development of spin-probe methods for these measurements overcomes this difficulty by virtue of the measurability of ESR spec-

Structure	Name	$t_{1/2}$(25°C)
—H	TEMPO	≤ 100 mSec
—OH	TEMPOL	≤ 100 mSec
$—NH_2$	TEMPAMINE	≤ 100 mSec
=O	TEMPONE	≤ 100 mSec
—COOH	TEMPCARBOXYLATE	≤ 100 mSec
$—CNH_2$ (C=O)	TEMPAMIDE	≤ 100 mSec
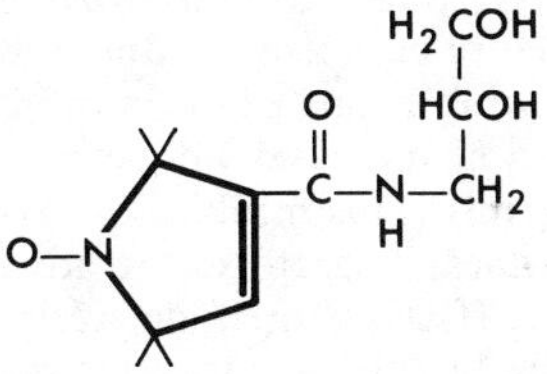	PYRGLYC	40 sec
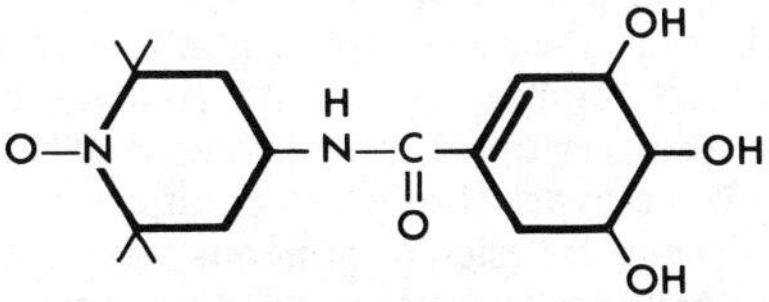	TASHIK	30 min
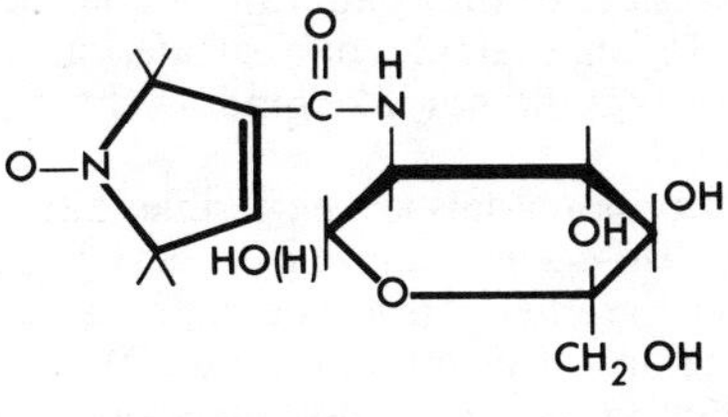	PYRGLUCOSAM	15 hrs
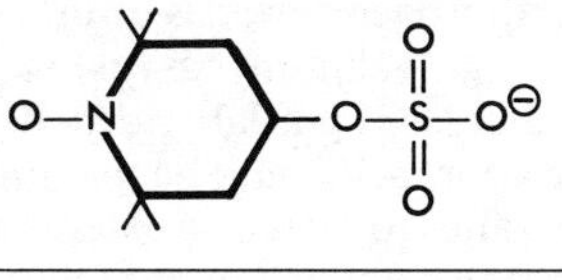	TEMPSULFATE	70 min
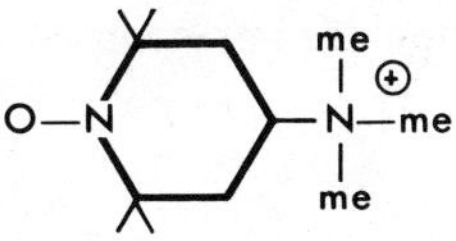	CAT 1	280 days

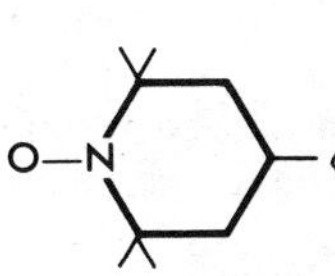

FIGURE 4. Half-times for full equilibration of nitroxide across membranes of S9 vesicles. Permeability of halobacterial vesicles to spin probes.

tral components because of bound ions so that this probable source of artifacts can be eliminated.

Two classes of permeable ions have thus far shown utility for ΔΨ measurements. These are lipophilic cation analogues of tetraphenyl phosphonium and the anion nitrophenyl phosphate esterified to 4-hydroxy-2,2,6,6-tetramethyl piperidine-*N*-oxyl (TEMPOL). These spin probes are schematically represented in FIGURE 4 along with the kinetics of their response to light-induced potentials in halobacterial envelope vesicles. All of the probes respond relatively slowly and thus have limited application for kinetic experiments.

FIGURE 5 shows substantial differences among the probes for uptake into the vesicles, implying that probe permeability imposes the rate-limiting constraint. The probe Kϕ was a generous gift from K. Hideg. The most rapidly accumulated probe, TϕE3, the ester of (3-carboxypropyl)triphenyl phosphonium bromide and TEMPOL, synthesized according to Cafiso and Hubbell,[9] also exhibits the greatest membrane binding. Not surprisingly, these lipid-soluble ions exert an inhibitory effect on ΔΨ development as the probe concentration increases above 50 μM (at a membrane concentration of about 10 mg protein/ml). This inhibitory effect is correlated with a binding of the probe to the membranes. Therefore, the advantage of rapid permeability of more hydrophobic ions is offset by their deleterious effect upon the functional integrity of membranes at the high concentrations required to give adequate signal-to-noise ratios of aqueous spectral lines.

Electrical potential measurements with spin-labeled phosphonium ions in illuminated halobacterial vesicles have shown that membrane-binding changes of the probe during energization have a substantial effect upon the calculated ΔΨ. In the usual calculation of ΔΨ it is assumed that all the hydrophobic ions that are removed from the bulk aqueous phase appear within the intravesicular aqueous phase. Analysis of ESR data has shown that much of this probe in fact becomes membrane bound and that the ratio of aqueous ions on the two sides of the membrane is considerably smaller than had been estimated previously. In one example, our calculation yielded a ΔΨ of 78 mV, where a conventional analysis that did not take binding charges into account gave a potential of 126 mV.[4]

At high salt concentrations, where surface potentials are negligible, alterations in the distribution of hydrophobic ions may also occur because of energy-linked dipole potentials, i.e., intramembrane electrical gradients (such as might arise from electron delocalizations into hydrophobic regions of membrane proteins). Thus far we have not attempted to define the contribution, if any, of boundary potentials to the membrane binding of hydrophobic ions. One approach towards estimating such effects might be to collapse ΔΨ with an ionophore, e.g., valinomycin (in the presence of potassium), to determine whether binding changes of spin-labeled hydrophobic ions occur upon membrane energization. Another approach could be to compare energy-linked responses of probes that bind to membranes to differing extents and thus would be expected to achieve the same transmembrane equilibrium distribution in response to ΔΨ, but different water-to-membrane partitionings in response to boundary potentials.

SURFACE POTENTIALS

Interfacial charge separations, and hence surface potential changes, may play a primary role in many membrane energization processes. Moreover, pH changes in aqueous domains will inevitably be accompanied by surface potential fluctuations as amino acid residues of membrane proteins lose and gain protons. Such surface po-

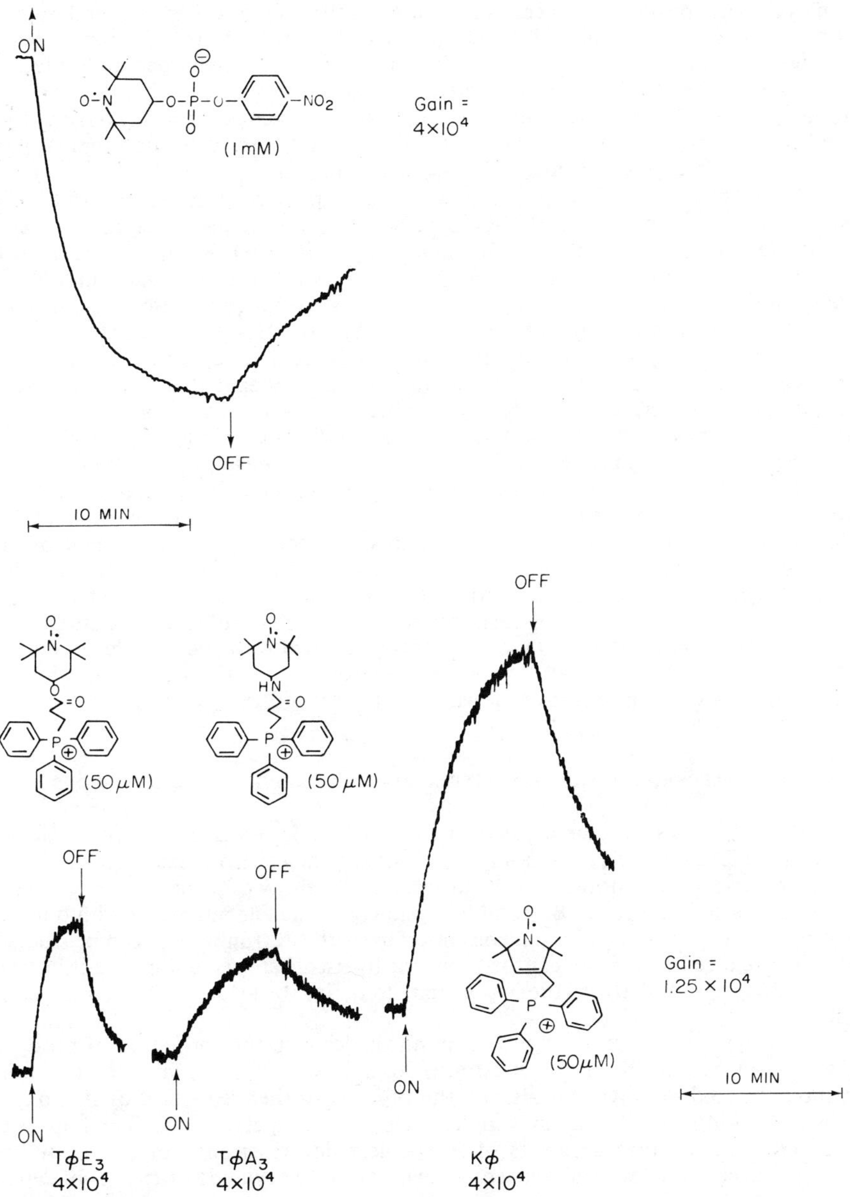

FIGURE 5. Kinetics of reequilibration of lipid-soluble ionic nitroxides across S9 vesicles as a function of illumination. 4 M NaCl, pH 7, 100 mM Na_3FeCN_6 for cations, 200 mM Na_3FeCN_6 for the anion.

tential fluctuations may have important consequences for bioenergetics, e.g., by modulating ion fluxes coupled to $\Delta\Psi$ or by affecting the substrate affinity for enzymes that use positively or negatively charged substrates at different membrane interfaces. Surface potentials will also be implicated in the response of many probes to energization. For all but the highest salt concentrations, any charged molecule with some membrane affinity, including virtually all of the probes in current use for measuring pH and electrical gradients, will respond to surface potentials. To correct for contributions of surface potentials to such probe responses it is important to have separate and unambiguous assays for surface potentials.

Long-chain, spin-labeled amphiphiles have proven effective for measuring membrane surface potentials.[10] These spin-labels are analogues of the familiar detergents SDS and CTAB (cetyltrimethyl ammonium bromide). Paramagnetic quenching techniques outlined in earlier sections have shown that the cationic amphiphiles, designated as CATn, permeate membranes very slowly (equilibration time is many hours in most membranes) and hence are suitable for side-specific measurements if loading procedures can be employed to selectively label the cytoplasmic membrane interface. On the other hand, the anionic probes, designated as ANn, permeate membranes considerably more rapidly (significant uptake occurs in a few minutes in most membranes) and hence are suitable for studies of average surface potential changes for both membrane interfaces (in single bilayer membrane vesicles or cells).

The surface potential assays rely entirely on discriminating between the clearly resolved spectral features of the probes in water and bound to membranes and thus do not require paramagnetic quenching agents. Furthermore, since a change of surface potentials leads to a reequilibration of probes across the aqueous interface without the requirement for transmembrane diffusion, these probes have the most rapid possible response kinetics of all the nitroxides being used for bioenergetic studies. Indeed, in measurements of surface potentials associated with proton release and uptake in halobacterial purple membranes, we found that the cationic probes responded to laser flash energization within 8 milliseconds.[11]

One-electron Reduction/Oxidation and Free Radical Reactions

For the studies described thus far spin reduction or destruction by cells is at best an annoyance. However, the study of nitroxide one-electron redox reactions can yield valuable information about the location and identity of membrane-bound electron donors and acceptors. By combining slowly permeable nitroxides with quenching agents, reactions at either aqueous interface with electron donors can be studied. Purified reduced nitroxides can be prepared by ascorbate reduction and thin-layer chromatography and afford the opportunity to define the location of electron acceptors.

Two types of processes can destroy nitroxides with the formation of products that cannot participate in one-electron redox reactions. One is their reaction with carbon-centered or thiyl radicals and the other is further reduction of hydroxylamines to form secondary amines under highly reducing conditions. The impact of such processes on measurements of reversible redox reactions can be assessed by treating broken cells with about ten equivalents of ferricyanide after energy-dependent reduction and oxidation assays have been completed. Such treatment will cause complete reoxidation of hydroxylamine derivatives of the nitroxides without affecting the other possible reaction products of the nitroxides.

Conclusions

Typical measurements of cell bioenergetics with probe techniques interpret data in terms of some single, average parameter and often neglect other processes that can affect the response of the probes. Spin labels provide detailed information about the partitioning of probes into specific membrane and aqueous domains and thus afford the opportunity to define more highly resolved energy-linked electrochemical potentials than have hitherto been obtained.

References

1. McConnell, H. M. 1976. *In* Spin Labeling. L. J. Berliner, Ed.: 525–560. Academic Press. New York.
2. Morrisett, J. D. 1976. *In* Spin Labeling. L. T. Berliner, Ed.: 274–338. Academic Press. New York.
3. Keith, A. D. & W. Snipes. 1974. Science **183**: 666–668.
4. Mehlhorn, R. J., P. Candau & L. Packer. 1982. Methods Enzymol. **88**: 751–762.
5. Quintanilha, A. T. & L. Packer. 1977. Proc. Natl. Acad. Sci. USA **74**: 570–574.
6. Gracetta, P. & J. C. Seidel. 1980. Biochemistry **19**: 33–39.
7. Quintanilha, A. T. & R. J. Mehlhorn. 1978. FEBS Lett. **91**: 104–108.
8. Mehlhorn, R. J. & I. Probst. 1982. Methods Enzymol. **88**: 334–344.
9. Cafiso, D. S. & W. L. Hubbell. 1978. J. Am. Chem. Soc. **17**: 187–195.
10. Mehlhorn, R. J. & L. Packer. 1979. Methods Enzymol. **56**: 515–526.
11. Carmeli, C., A. T. Quintanilha & L. Packer. 1980. Proc. Natl. Acad. Sci. USA **77**: 4707–4711.

INDEX OF CONTRIBUTORS

SUBJECT INDEX